A

# PRACTICAL TREATISE

ON

# RAILWAY CURVES

AND

# LOCATION,

## FOR YOUNG ENGINEERS.

CONTAINING A FULL DESCRIPTION OF THE INSTRUMENTS, THE MANNER OF ADJUSTING THEM, AND THE METHODS OF PROCEEDING IN THE FIELD,—NEW AND SIMPLE FORMULÆ FOR COMPOUND AND REVERSE CURVING,—RULES FOR CALCULATING EXCAVATION AND EMBANKMENT,—STAKING OUT WORK, &c., TOGETHER WITH TABLES OF NATURAL SINES AND TANGENTS, RADII, CHORDS, ORDINATES, AND OTHERS OF GENERAL USE IN THE PROFESSION.

BY

WILLIAM F. SHUNK,

CIVIL ENGINEER.

PHILADELPHIA:
HENRY CAREY BAIRD,
INDUSTRIAL PUBLISHER,
No. 406 WALNUT STREET.
1869.

# PREFACE.

The located line for railway is a series of curves and straight lines, or tangents. These are first plotted to a large scale from data gathered on preliminary survey. It is therefore desirable that all explorations should be made with extreme care, as upon their correctness depend, in no small degree, the labour and time required in location. It were better for accuracy that all angles should be made and recorded from the plates, and the needle used only as a test, or check. Good chaining is indispensable. Great attention, too, should be given to the proper use of the slope instrument. By these means a working map can be constructed in the office upon which the proposed location, grade lines, &c., may be traced with tolerable resemblance to fact. Still many errors attach to both data and map, and these, together with the unexpected obstacles encountered in the field, require ready knowledge of the means for overcoming them.

It has been my design to present this knowledge to my younger fellows in the profession. I have endeavoured do it lucidly and concisely—without supposing unusual cases—without prolix proof or complex figuring. The problems given are of frequent occurrence, and the tables appended will be found useful and correct.

To Strickland Kneass, a gentleman whose professional abilities are well known, I return thanks for valuable assistance. I would likewise make my acknowledgments, for useful suggestions, to Charles Delisle, an engineer of high mathematical attainment.

I am aware that much more might have been said—much more suggested—on the subject of location; but a field book being the object, the compact plan precluded any extensive essay.

If the work with brevity combines clearness, and is comprehensive withal, it is the work intended.

W. F. Shunk.

# CONTENTS.



1*

## TABLES.

# EXPLANATIONS.

---

All railway curves are parts of circles. They are designated *generally* from their character as simple, compound, or reverse; and *specifically* from the central angle subtended by a chord of 100 feet at the circumference, this being the length of the chain in common use. It is found that the circle described with radius of 5730 feet has a circumference of 36,000 feet. Since there are 360° in the circle, the central angle subtended by a chord of 100 feet is, in this case, equal to 1°, and the curve is named a *one degree* curve. So likewise in a circle with radius of 2865 feet, half of 5730, the central angle corresponding to the chord 100 is 2°; the curve is then called a *two degree* curve.

The beginning of a curve is called the point of curvature, or simply the P. C., and its termination the point of tangent, marked P. T.

A compound curve is composed of two curves of different radii, turning in the same direction, and having a common tangent at their point of meeting. This point is called the point of compound curvature, or P. C. C.

A reverse curve is composed of two curves turning in different directions, and having a common tangent at their point of meeting, which latter is named the point of reversed curvature, or the P. R. C.

All sines and tangents made use of in this work are from the table at the end of the volume. For calculating curves it is not necessary to use more than four decimals.

*A Bench* is a shoulder hewn with the axe on the buttressed base of a tree, and so shaped at the top as to afford footing to the rod. The tree is blazed and the elevation of the bench marked on it with red chalk. Benches serve as permanent reference points to the level. They are placed, where it is possible, about one thousand feet apart.

*Points.* The operation termed pointing is the fact of putting a peg firmly into the ground, and of driving in its top a tack, or making thereon an indentation whose place is indicated by cross keel marks, directly in the line of collimation of the transit. Thus true lines are traced on the ground, and angles measured accurately. When the transit is set over a point it is so posited that the plumb hangs immediately above the tack head. If the head plate of the tripod be much inclined the plumb should be examined after levelling the instrument, as that operation disturbs it to some extent.

*Stations.* The line of a survey is marked on the ground at regular intervals, by stakes two feet in length, blazed, and numbered from 0 up in arithmetical progression. These stakes are named stations. On exploration they are commonly placed two hundred, and on location, one hundred feet apart.

It is customary, when locating, to drive pegs even with the surface along the true line, and to place the stations a couple of feet to the right, numbers facing in, to show their position. The pegs are less liable to disturbance from frost, animals, &c.

In locating for construction stakes are driven on sharp curves at intervals of 50, sometimes 25 feet.

*The Chain* in general use for railway surveys is made of soft iron. It is 100 feet long, and divided into 100 links, each one foot in length. At every tenth link is attached a brass drop, toothed so as to indicate its distance from the end. It presents the advantages of durability, accuracy, and expedition.

# A PRACTICAL TREATISE

ON

# RAILWAY CURVES AND LOCATION.

---

## ARTICLE I.

### OF THE INSTRUMENTS.

#### THE LEVEL.

THE level is an instrument used in ascertaining the undulations of the ground along the line of a survey, and of measuring these irregularities accurately in reference to an assumed base called the datum. It consists mainly of the telescope *k i*, the spirit-level and its encasement *b*, the Y's *c c*, the rectangular bar *o d*, the axis *e*, the plates and levelling screws *f m*, and the tripod *g*.

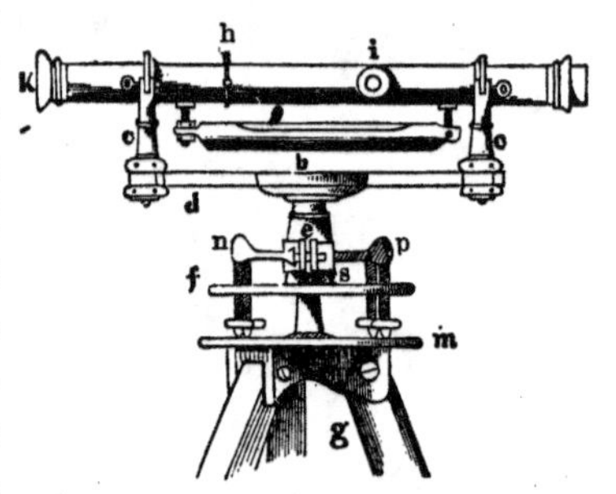


In the tube of the telescope at *h*, and at right angles to its axis, is placed a flat ring, called the diaphragm. To this ring the cross-hairs are attached—two delicate spider lines stretched over it vertically and horizontally, and intersecting at the centre. It is held in position by four slightly movable screws, which pierce the tube in the direction of the "cross-hairs." *i* is a milled head for adjusting the focus of the object glass, and *k* an inserted tube, containing several lenses, which may be moved out or in so as to make the spider-lines distinctly visible.

A straight line looked along from the eye glass at *k* through the intersection of the cross-hairs is the *line of sight*, technically named the *line of collimation.*

The immediate supports of the telescope are called the Y's, from their resemblance to that letter. If a small arch were sprung between the two legs of the Y it would give a good idea of the clasping pieces which hold the telescope in place. They are jointed to one leg and secured to the other by pins which may be withdrawn and the pieces turned back in order to remove the telescope, or change it end for end.

The Y's are attached at right angles to the bar *d*, which, again, is connected firmly at right angles with the hollow axis *e*. This latter fits closely over and is revolvable horizontally around a *solid* axis *s*, which, passing through the plate *f*, is secured to the head of the tripod by means of a loose ball-and-socket joint. The plate *f* has four levelling screws inserted in it; with these the instrument may be brought to a horizontal position even when the lower plate is considerably inclined.

One of the Y's is movable for a short space up or down by means of the capstan-head screw *o*. The spirit-level is likewise movable both vertically and laterally by means of screws at either end.

*n* is a clamp screw, and *p* a tangent-screw for slightly turning the telescope in a horizontal direction.

*To adjust the Level.*

*First. To make the line of collimation coincide with the axis of the telescope.*

Set the instrument firmly, and direct the telescope toward some distant, distinct object, such as a nail-head. Clamp fast, and with tangent-screw fix the line of collimation upon the object accurately. Revolve the telescope half way round in the Y's, *i. e.* until the bubble is above it, and if the horizontal spider-line still covers the point, it requires no adjustment. If it does not, reduce the error one-half by means of the diaphragm screws, and complete the reduction with the capstan-head screw. Revolve the telescope round to its first position, and if the horizontal line and point do not then coincide, repeat the operation until they do, in any position of the telescope. In similar wise the vertical hair may be adjusted, when the line of collimation should cover the point through an entire revolution of the telescope.

Great care should be taken in this as well as in all other adjustments of cross-hairs, that the opposite screw of the diaphragm be loosened before tightening its fellow, or injury to the instrument must result.

*Second. To make the axis of the spirit-level parallel to the line of collimation.*

With levelling screws bring the bubble to the middle of its tube, reverse the telescope in its Y's, and if the bubble does not then stand in the middle correct one-half the deviation with the screw at the left end of the bubble-case, and the other half with the capstan-head screw. Again reverse the telescope in its Y's, and, if necessary, repeat the operation.

Now revolve the telescope a short distance in its Y's, so as to bring the spirit-level to one side of its lowest position. If the bubble deviates from the middle, correct the error with the lateral screws at the right end of the bubble-case, and examine the previous adjustment before lifting the instrument.

*Third. To bring the line of collimation parallel to the bar.*

Turn the telescope until it stands directly over two of the levelling screws, and with them bring the bubble to the middle of the tube. Then revolve the telescope horizontally until it stands over the same screws, changed end for end. If the bubble does not still stand in the middle of the tube, correct one-half the deviation with the capstan-head, and one-half with the levelling screws.

Place the telescope over the other levelling screws and proceed in a similar manner, and continue the corrections until the bubble stands without varying during an entire revolution of the instrument upon its axis.

This completes the adjustment of the level.

## THE ROD.

The rod used in levelling consists of a staff and a target, which latter is so attached to the staff as to be movable along it from end to end. The rod is commonly seven feet long, but, being composed of two rectangular pieces fitted together by means of a sliding groove, it can be extended to nearly double that length. It is graduated to feet and tenths of a foot. The target is a circle of wood or iron, usually four-tenths in diameter, and divided into quadrantal sectors by a horizontal and vertical line which intersect at its centre. The sectors are painted alternately red and white, so that their dividing lines are visible at a considerable distance. On the back of the target, where it meets the graduated side of the rod, is fixed a chamfered brass edging, whereon the space of one-tenth is graven from the centre down. This is subdivided into ten spaces marking hundredths, and these latter divided into halves, so that the height of the middle of the target above the base of the rod may be accurately read to within ·005 of a foot.

There is a similar graduated tenth on the standing part of the rod, to be used for high sights when the sliding groove comes into play.

Both target and rod are provided with clamp screws.

## LEVELLING.

The operation technically called levelling is performed thus:—

Suppose *a* the starting point, or zero, in reference to which all the inequalities of the surface along the line of survey are measured, as at the points *c*, *e*, *f*. The horizontal line *a f* is called the *datum* line. This is arbitrarily

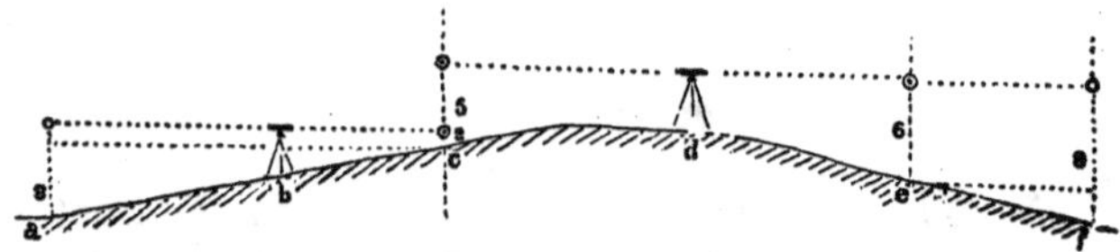

assumed. It may be considered, for example, at any dis tance *above* the point *a*, and the irregularities of the ground measured from an imaginary level line in ether; but for convenience of figuring, and other politic reasons, it is customary in seaport towns to take high tide as datum. Inland, the summer surface of the nearest stream, or, when commencing on a ridge, the highest neighbouring knoll is assumed.

Well! suppose *a* to be zero, and the instrument, for instance, set and levelled at *b*. Stand the rod at *a*, and slide the target up until its cross-lines are covered by the cross-hairs in the telescope; *i. e.*, until the line of collimation coincides with the centre of the target. The leveller directs the movements of the target by raising or lowering his hand. A circular motion of the hand signifies "make fast." The bubble should always be examined before the rod is taken down, and the latter should be read twice, or, if convenient, shown to the leveller, in order to guard against mistake. If in this case it reads 8 feet, the height of the instrument is then 8 feet above *a*. To find the elevation of *c* above *a*, take the rod thither and lower the target until coincidence results as before. If the rod reads 2 feet, of course *c* is 8 — 2 = 6 feet above *a*.

If it is necessary to lift the instrument here, a small peg is driven at *c* before sighting to that point, to insure firm footing for the rod. Sighting back from the new position, *d*, the rod reads 5 feet; then 5 + 6, the elevation of *c* above *a*, = 11, the height of the telescope at *d* above *a*. If at *e* the reading is 6, the elevation of that point is 11 — 6 = 5, and if at *f* the reading is 8 the elevation of that point in reference to *a* is 11 — 8 = 3, marked + 3.

The rule, therefore, in levelling is, at each new stand of the instrument, to add the reading of the rod sighted back at, to the discovered elevation of the point at which the rod stands, for the height of the instrument; and to *subtract* from this height the reading of the rod at any points observed from the new position in order to find the elevation of those points. The above is noted in the field-book as follows:

| Station. | Rod. | Height of Instrument. | Total, or Elevation. |
|---|---|---|---|
| *a* | 8·00 | 8·00 | 00 |
| *c* | 2·00 | | + 6.00 |
| | 5·00 | 11·00 | |
| *e* | 6·00 | | + 5·00 |
| *f* | 8·00 | | + 3·00 |

The advantages of this method of levelling over the old system of backsights and foresights are, that it affords readier facilities for testing the correctness of the work, and it may be carried on more rapidly. By the old plan each sight at the rod was linked with that which preceded it, and added one more to a continuous calculation in which a single error affected all the following work. Here, however, if haste is required, the calculation of the intermediate sights or "cuttings" may be omitted entirely while in the field, the reading of the rod only being set down; the "totals" may be worked from peg to peg, and the liabilities to mistake thus decreased about eighty per cent.

## OF THE TRANSIT.

The transit is an instrument for measuring horizontal angles. It consists of the telescope *a c*, the Y's *d*, the compass-box, &c., *e g*, and the axis *k*. The telescope is furnished like that of the level, and the instrument is similarly fitted to its tripod. The telescope revolves in a vertical circle, and is attached to the Y's by means of a transverse axis whose extremities turn in smooth journals at the head of the Y's. The body of the instrument at *f* contains a magnetic needle, with its usual circular surrounding, graduated to degrees and quarter degrees. The flooring of this box has, on one side, an opening with chamfered edge upon which the vernier is engraved. This latter, together with the telescope, Y's, and all the upper part of the instrument, is made to revolve by means of the screw *h*, upon a solid plate beneath, which is likewise graduated from 0 to 180° each way. Thus angles may be measured

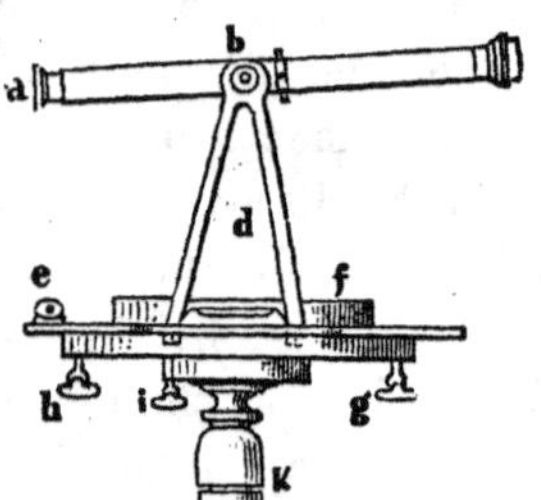

accurately without using the needle at all. It need be regarded merely as a check. *g* is a clamp screw for securing the plates together, and *i* a screw for fastening the needle so as to prevent its vibrations while the instrument is being carried from place to place. A plumb is suspended from the axis of the transit, by means of which its centre may be placed over a point on the ground.

## THE VERNIER.

The vernier, in the transit, is a graduated index which serves to subdivide the divisions of the graduated arc on the lower plate. There are many varieties of the vernier, but familiarity with one renders easy the acquaintance with all, since the same general principle is pervading.

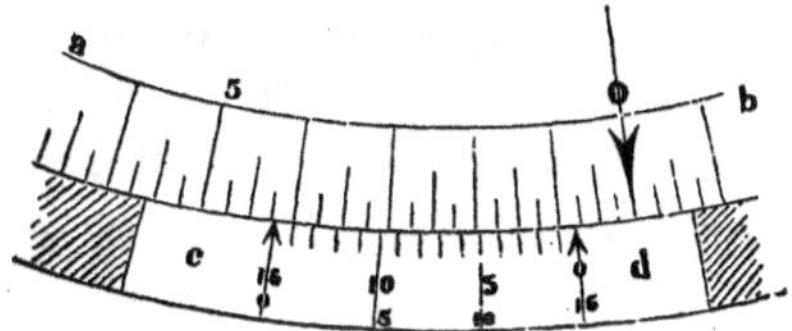

The figure represents a common form. Let *a b* be part of any graduated arc, and *c d* the vernier. It will be observed that the degrees on the limb are divided into spaces of 15′ each. Now if the vernier be made equal in length to fourteen of those spaces, and be further divided into *fifteen equal parts*, it is evident that each of these parts will contain 14′.

Then, if 0 of the vernier coincides with any division of the limb, the first line of the vernier to the left will be just one minute behind the first line of the limb to the left; the *second* vernier line *two* minutes behind the second limb line, and so on; so that if the vernier be moved to the left over the space of 15′ on the limb, the lines from 0 to 15 of the vernier would coincide successively with lines of the limb, and thus any angle may be read accurately to minutes.

The vernier in the figure reads 48′ to the left. A vernier graduated decimally is much more convenient on railway locations than those with the common graduation to minutes. This is principally on account of its adaptedness to running in curves when the 100 feet chain is used. The work can be done with more ease and rapidity. One objection to it is that the tables in general use are calculated for degrees and minutes.

## TO ADJUST THE TRANSIT.

Place the instrument firmly at *a*, level it, clamp all fast, and with tangent-screw set the cross-hairs on the point *b*, at any convenient distance. Reverse the telescope on its axis, and fix another point in the opposite direction,

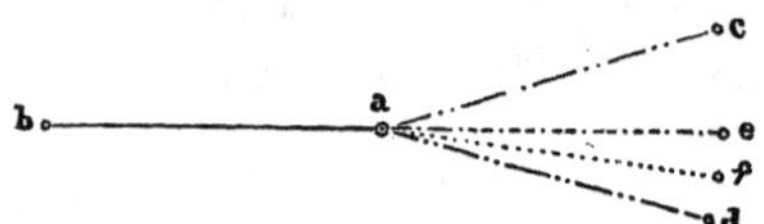


as nearly as possible equidistant from *a*. Now loose the lower clamp and revolve the entire upper part of the instrument half way round on its axis. Clamp fast, and having brought the cross-hairs again to coincide with *b*, reverse the telescope. If the sight strikes as before, the instrument is in adjustment. If not, place another point, *d*, where it does strike, and suppose *c* to be the point previously fixed: the point *e*, midway between *d* and *c*, is then in the straight line. With the adjusting pin carefully place the vertical cross-hair upon *f*, distant from *d* one-quarter of the space *d c*—with tangent-screw set it on *e*, and reverse the telescope. If the points have been correctly placed, and the hair properly moved, the sight will strike *b*, and the adjustment is complete.

After finishing this adjustment, the telescope may still not revolve truly in the meridian. This inaccuracy there is no method of removing in the field. It should be sent to an instrument-maker for repairs.

---

## ARTICLE II.

### PRELIMINARY PROPOSITIONS.

1. *In any circle* the angle $o\,c\,f$ at the centre, subtended by the chord $o\,f$, is double the angle $o\,a\,f$, at any part of the circumference on the same side of the chord.

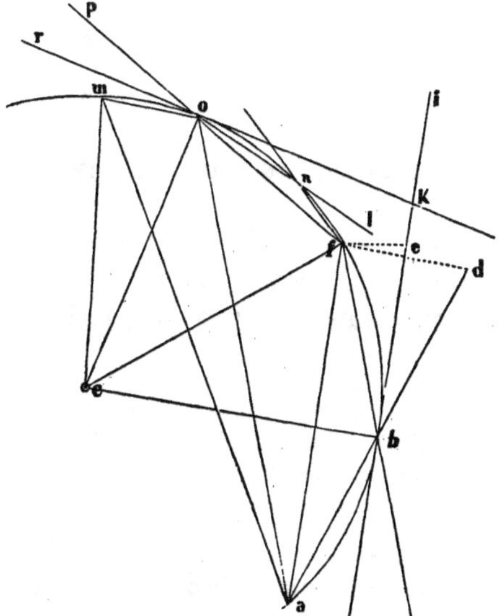


2. *The angle* $f\,b\,e$, *formed* by any chord $f\,b$, with a tangent at either extremity, is called a *tangential* angle.

and is equal to half the angle *f c b* at the centre, or is equal to the angle *f a b* at the circumference.

3. *The exterior angle d b f*, formed at the circumference by the two equal-chords *a b*, *b f*, is called a *deflexion* angle, and is equal to the central angle *b c f*, or double the tangential angle *e b f*. *d f* is called the *deflexion distance*, and *e f* the *tangential distance*.

4. *The exterior angle p o m* of two unequal chords, is equal to the sum of their tangential angles, or half the sum of their central angles.

5. *The exterior angle i k o*, formed by tangents, is equal to the central angle *b c o*, subtended by the chord which connects their points of contact with the curve.

---

## ARTICLE III.

### TO AVOID AN OBSTACLE IN THE LINE OF TANGENT.

A GLANCE at the figure will show, that having deflected to *c*, and placed the instrument at that point, the angle *h c d* must be made equal to twice *d b c*, and the distance

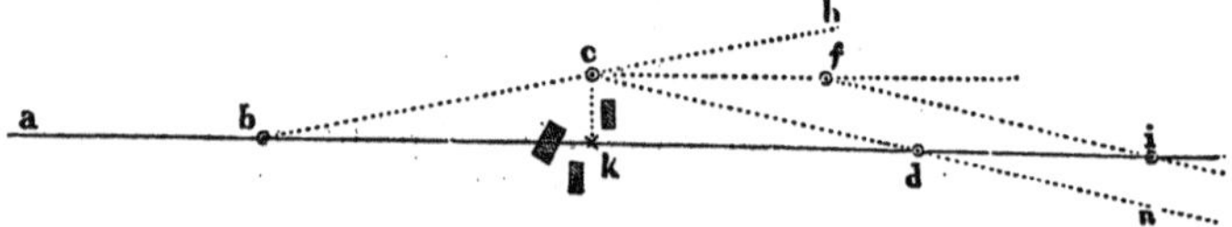


*c d* equal to the distance *b c*. Still another deflection at *d*, equal to the original angle turned, is necessary in order to sight again along the tangent.

Should the obstruction be continuous a parallel line may be run, as from *c* to *f*, by deflecting at *c* an angle *equal* to that at *b*, and at *f*, repeating the deflection in order to strike tangent.

If the angle *d b c* exceeds 4°, and the distance *b c* is greater than 200 feet,—or even with an angle of 2½°, should the distance be greater than 300 feet,—*b c* will differ sensibly in length from *b k*, and a calculation of the latter becomes necessary. To effect this, multiply the natural cosine of the angle *k b c* by *b c*. This result *doubled* will give *b k d*, the length proper along tangent.

Thus:—Suppose *k b c*, the angle deflected, to have been 5°, and the distance *b c* 340 feet. Then ·9962, the natural cosine of 5°, multiplied by 340, gives 338·7 for the distance *b k*. Double this makes *b k d* = 677·4, and shows a difference of 2·6 feet between *b d* and *b c d*.

---

## ARTICLE IV.

### SHOULD THE OBSTRUCTION LIE ON THE OPPOSITE BANK OF A STREAM,

AND it is desirable on any account not to run the line from *d*, corresponding to *a d̄*, set the instrument at *a*, in tangent, and deflect clear of the obstacle to *d*. *Point* at *d*, deflect to *e*, and point also there—marking the angles *c a d*, *d a e*. Chain the base *d e*, and placing the transit at *e*, measure the angle *d e a*. Data are thus obtained sufficient for the calculation of the line *d a*. The object now is to find the point *c* and the angle *d c a*. The angle *a d e* subtracted from 180° will supply the

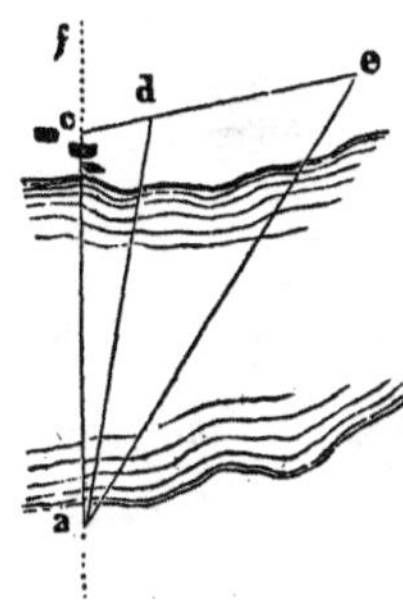

angle $c\,d\,a$, so that in the smaller triangle we have obtained two angles and their included side. The distance $c\,d$, and angle $d\,c\,a$ readily follow. The transit standing at $e$, $c$ is placed, of course, in the prolongation of the base $d\,e$, and the distance $c\,d$ is carefully set off with the rod. Moving the instrument to $c$, and turning the angle $e\,c\,f = 180° - d\,c\,a$, we are again in tangent.

*Example.*—Let $c\,a\,d = 6°$, $d\,a\,e = 35°$, $d\,e\,a = 42°$, and the base $d\,e = 200$ feet. Then in the triangle $d\,e\,a$ we have

Nat. sine $d\,a\,e = (\cdot 5736)$ : nat. sine $d\,e\,a = (\cdot 6691)$ :: $d\,e = (200) : d\,a$,

$$\text{Wherefore,} \quad d\,a = \frac{\cdot 6691 \times 200}{\cdot 5736} = 233\cdot 3 \text{ feet.}$$

Again, the angle $c\,d\,a = 77°$. The angle $d\,a\,c$ being $= 6°$, $a\,c\,d$ is consequently $= 97°$, and in the small triangle we have, Nat. sin. $a\,c\,d = (\cdot 9925)$ : nat. sin. $c\,a\,d = (\cdot 1045)$ :: $d\,a = (233\cdot 3) : c\,d$.

$$\text{Therefore } c\,d = \frac{\cdot 1045 \times 233\cdot 3}{\cdot 9925} = 24\cdot 564 \text{ feet, and } d\,c\,f$$

$180° - 97° = 83°$.

NOTE.—A common and convenient plan for triangulating a creek is as per figure. Set the instrument at $b$, fix a point $d$ on the opposite shore, and making $d\,b\,c$ a right angle, place $c$ at any convenient distance. Now move to $c$, sight to $d$, and making $d\,c\,a$ a right angle also, fix $a$, in the same line with $b$ and $d$. $a$, $c$, and $d$ are points in the circumference of a circle whose diameter is $a\,d$, $a\,b : b\,c :: b\,c : b\,d$, and therefore $b\,d = \frac{bc^2}{ab}$.

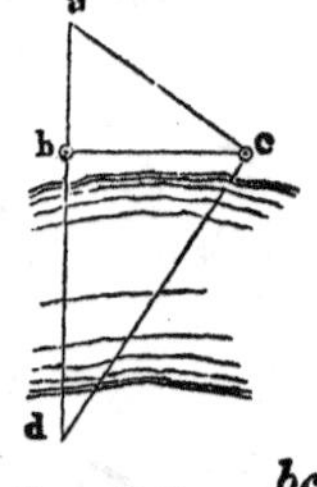

## ARTICLE V.

HAVING GIVEN THE ANGLE *e d b*, FORMED BY THE INTERSECTION OF TWO STRAIGHT LINES, IT IS REQUIRED TO FIND THE POINT *a* OR *b*, AT WHICH TO COMMENCE A CURVE OF GIVEN RADIUS.

Draw the bisecting line *d c*. Then the angle *d c a* = half the angle *a c b* or its equal *e d b*; and in the triangle *d c a*, the angle *d a c* being a right angle, we have

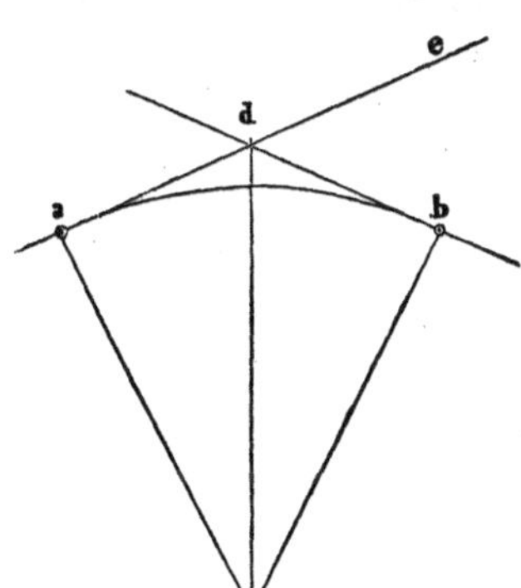


Rad. of 1 : Nat. Tang. *d c a* : Rad. *c a* : *a d*. Therefore *a d* = Nat. Tang. *d c a* × Rad. *c a*.

*Example* 1.—Let *e d b* = 48° and *a c* = 1460 feet. Here half the angle *e d b* or *a c b* = 24°, the Nat. Tang. of which is ·4452; and multiplying by Rad. 1460, we have 650 feet for the length of *a d* or *d b*, the tangents.

2.—*If a d be given and radius required*,—

$$\text{Rad.} = \frac{a\,d}{\text{Nat. Tang. } a\,c\,d} = \frac{650}{\cdot 4452} = 1460.$$

The following rules are approximate, and sufficiently correct for all purposes of location.

To find the degree of curvature of *a b* divide 5730 by the radius in feet; and to find the length of the curve in feet divide the angle *a c b* (after reducing minutes to hundredths) by the degree of curvature—the chord in each case being 100 feet in length.

## ARTICLE VI.

### TO TRACE A CURVE WITH TRANSIT AND CHAIN.

THE degree of curvature and the angle to be turned are known. If the latter is expressed in degrees and minutes, reduce the minutes to hundredths, since the 100 feet chain is used, and divide the whole angle by the degree of curvature. The quotient will be the length of the curve in feet, and the P. T. is at once ascertained.

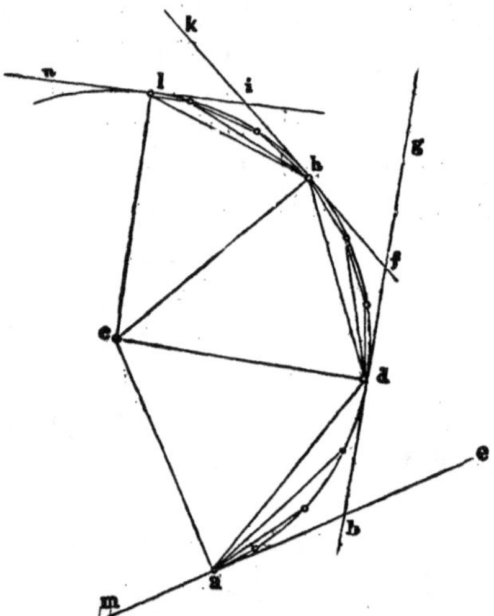


Let *m a* be the tangent, and *a* the P. C. Place the transit at *a*, index reading 0, and direct the sight along the tangent *m a e*. The first deflection will be half the central angle subtended by the chord used, and all the stakes put in from *a* will be fixed by similar tangential deflections. (*Prelim. Prop.* 1.)

When the point *d* is reached, the angle *d a b*, shown on the index, will be half the angle *d b e*, or its equal *a c d*, at the centre. Move the instrument to *d*, sight back to *a*, and turn to *double the index angle.* The telescope is now directed along the tangent *b d g*, and the angle $d b e = a c d = d a b + a d b$, reads on the index. Note this angle in the column of tangents opposite station *d*. Continue the curve from this new position, precisely as was done at *a*, and set the point *h*. Move to *h*, see that the vernier has not been disturbed, and sight back to *d*. The index now shows the angle $(d b e + h d g)$, and the object is to turn the angle *d h f*, i. e. *repeat* the angle *f d h*, as was done before at *d*, and, at the same time, have the whole angle $(d b e + h f g)$ indicated on the plate. To effect this, merely add this angle *f h d* to the present reading. It will be found simpler, in practice, to double the entire angle thus far turned, and subtract from the product the last tangent, viz. *d b e*. The vernier, turned to this resultant angle, will put the telescope in tangent line to *h*. *And so on.*

*Example.*—"At sta. 24 + 50 commence a 4° curve to the left for 35° 12′." Suppose this a required duty. First, reducing minutes to hundredths, we have 35°·20, which, divided by 4°, gives 880 feet for the length of the curve. Adding 880 to 24 + 50 it is at once seen that sta. 33 + 30 is the P. T.

Let *a* be the P. C., = 24 + 50. Now the *deflexion angle* being 4°, the *tangential* angle is 2°, with a chord of 100 feet. With a chord of 50 feet, therefore, the tangential angle is 1°, and this deflexion from tangent *m a e* fixes station 25. A deflexion from this latter point of 2°, the chord being 100 feet, fixes station 26. *And so on.*

When you have fixed the point *d*, = sta. 28, the index reads 7°. Move up to station 28, sight back to the P. C., and turn the index to 14°. This throws you on tangent

Proceed as before, with the 2° deflexions, to sta. 31, = *h*. Move up, and sight back to sta. 28. The index now reads 20°. Multiplying by 2, and subtracting the last tangent, we have the reading of the tangent at *h* = 26°: we have turned 26° of the curve. Continue as before. After putting in sta. 33, to find the deflexion which shall fix the P. T., 33 + 30, say, as 100 feet : 30 feet : : 2° : the required deflexion, = 36′. We may here remark the great convenience of an instrument graduated to *hundredths* of a degree instead of *sixtieths*. In the present example it would be seen immediately that the tangential angle for 100 feet being 2°, for 1 foot it would be 2 *hundredths* of a degree, and for 30 feet it would be 60 hundredths.

Well! when the P. T., = 33 + 30, is fixed, the index reads 30° 36′. Move up, see that the vernier has not been disturbed, and sight back to sta. 31. Now twice the index reading, minus the last tangent, = 61° 12′ — 26°, = 35° 12′, the present tangent, which is the final tangent, which finishes the curve.

The advantage of this manner of running a curve is that the instrument shows at a glance the work done, and therefore errors may be detected with greater facility. By comparing at the P. T. the total index angle with the distance run, the work is tested at once.

The above is recorded in the field book as follows:—

| Station. | Distance. | Deflex'n. | Index. | Tangent. | Course. | Mag. Course. | Remarks. |
|---|---|---|---|---|---|---|---|
| 23 ○ | 100ft. | | | | N. 20°·00 W. | N. 20° 05 W. | |
| 24 | 50 | | | | | | |
| P. C. + 50 ○ | 50 | 1°·00 | 1°·00 | | | | At sta. 24 + 50 commence a 4° curve to the L. for 35° 12′. |
| 25 | 100 | 2·00 | 3·00 | | | | |
| 26 | 100 | 2·00 | 5·00 | | | | |
| 27 | 100 | 2·00 | 7·00 | | | | |
| 28 ○ | 100 | 2·00 | 16·00 | 14°·00 | | N. 34° 07′ W. | |
| 29 | 100 | 2·00 | 18·00 | | | | |
| 30 | 100 | 2·00 | 20·00 | | | | |
| 31 ○ | 100 | 2·00 | 28·00 | 28·00 | | | |
| 32 | 100 | 2·00 | 30·00 | | | | |
| 33 | 30 | 0·36′ | 30·36′ | | | | |
| P. T. + 30 ○ | 70 | | | 35·12′ | N. 55°·12 W. | N. 55° 18′ W. | |
| 34 | 100 | | | | | | |

In running compound and reversed curves the operation is quite as simple as the foregoing. A point is fixed at the P. C. C., or P. R. C., and turning into tangent at that point, the second curve is traced from this *tangent*, without regard to what precedes. In reverse curving, it is a good plan to adjust the index in such manner at the P. R. C., that when we turn into tangent it will read 0.

This saves troublesome work, and it is advisable moreover to show in the field-book the contained angle *of each curve*, as well as the test of the two tangents with the magnetic course.

---

## ARTICLE VII.

### TO TRIANGULATE ON A CURVE.

SET the transit at *a*, and, as usual, sight back, and turn into tangent. Estimate the distance to the farther bank —do it liberally—and make a deflexion around the curve, corresponding to your estimated distance. Fix a point *b* in this line. Measure any convenient angle, *b a c*, and set

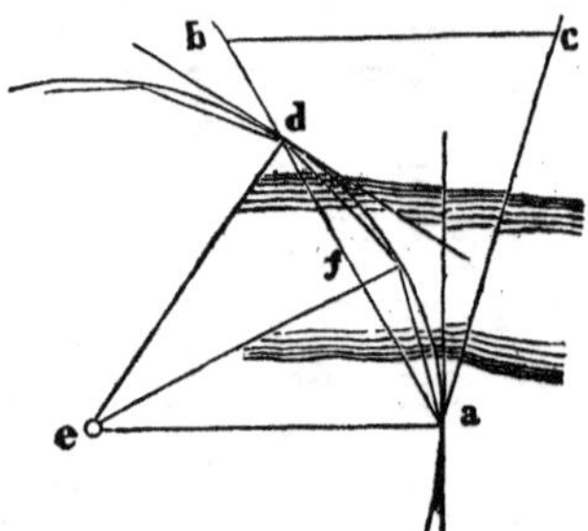


the point *c*. Move to *b*, measure the base *b c*, the angle *a b c*, and, before lifting the instrument calculate the line *b a*. If the angle turned from tangent to *d* exceeds 4°, and the distance is greater than 200 feet, the chord *a d* must also be calculated, as per example, and the difference between this and *b a* will be the distance *b d* to the point *d*, in the curve, which can be fixed from *b*.

Should $b$ fall between $d$ and $a$ the operation is analogous.

*Example.*—Let $a$ be a point in a 6° curve. Having set the transit, and turned into tangent, the distance to the farther verge is estimated 400 feet. The tangential angle for 100 feet is 3°, and to fix $d$, 400 feet distant, is consequently 12°. Deflect this angle, fix a point in line, and complete the triangulation, as previously illustrated in Art. IV., p. 22. Suppose $a\,b$ found equal to 472 feet. Now the tangential angle to $d$ = half the central angle, $= f\,e\,a$, = 12°; and to find the length of the chord $a\,d$, we have, in the triangle $e\,f\,a$,

Rad. : Sin. $a\,e\,f$ : : $e\,a$ : $a\,f$, that is

Rad. of 1 : Nat. Sin. 12° : : 955·4, the Rad. of the 6° curve : half the chord required. Wherefore $a\,d$ = twice the Nat. Sin. 12° × 955·4 = ·2079 × 2 × 955·4 = 397·25 feet. Subtracting this from 472, we have the distance, 74·75 feet, back to the point in the curve. Move the instrument to $d$, set the index at 12°, sight back to $a$, and turning to 24°, the telescope is in tangent. A deflexion of 3° will fix the next station.

NOTE.—In this case, if preferred, a third proportional might be formed with the chord of crossing, as shown in the note to Art. IV.

## ARTICLE VIII.

### TO CHANGE THE ORIGIN OF ANY CURVE, SO THAT IT SHALL TERMINATE IN A TANGENT PARALLEL TO A GIVEN TANGENT.

LET $d f$ be the located curve, terminating in a tangent $f k$, and the nature of the ground requires that it should terminate in the tangent $e i$, parallel to $f k$. At $f$, the telescope being directed along the tangent $f k$, turn to the

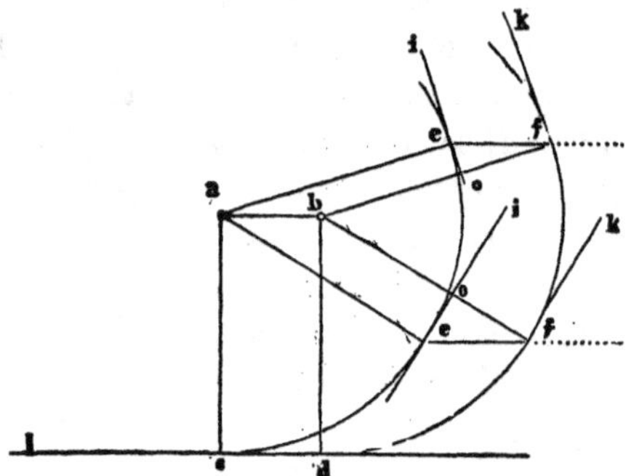


right an angle equal to the central angle $d b f$, previously turned to the left on the curve. This will direct the telescope along $e f$, parallel to $d l$. Measure $e f$, and go back on the tangent, $d l$, a distance, $c d$, equal to it. The curve, retraced from $c$ and consuming the same angle, will terminate tangentially in $e i$. An example in this case is not necessary.

## ARTICLE IX.

TO CHANGE A P. C. C. SO THAT THE SECOND CURVE SHALL TERMINATE IN A TANGENT PARALLEL TO A GIVEN TANGENT.

LET $a\ b\ d$ be the compound curve, located and terminating in the tangent $d\ h$. Continue the larger curve to $e$, and from $e$, with rădius $e\ l = k\ b$, describe the curve $e f$, terminating tangentially in $f\ g$, parallel to $d\ h$. From $c$, the centre of the larger curve, let fall upon $f\ g$ the perpen-

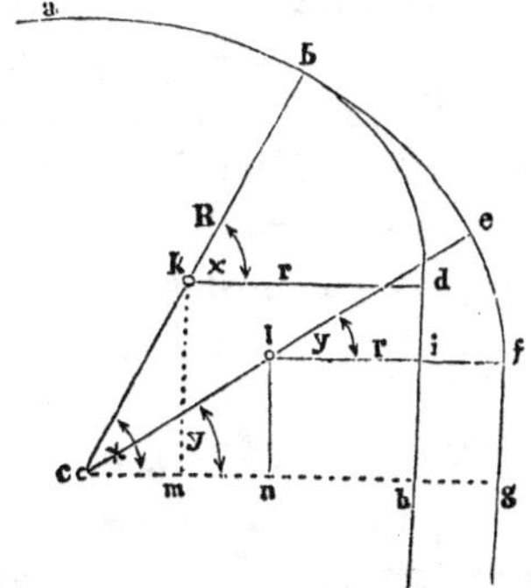


dicular $c\ g$, and fill up the figure as above. Call the radii respectively R and $r$, the angle $b\ k\ d$, or its equal, $k\ c\ m$, $x$, and the angle $e\ l f$, or its equal, $l\ c\ n$, $y$. Let the distance $i f$, or $h\ g$, be named D. Now the line $c\ g$ is made up of the lines $c\ m + m\ h + h\ g$, i. e., $c\ g =$ cosin. $x$ $(R - r) + r +$ D. $c\ g$ is also made up of the lines $c\ n + n\ g$, i. e., $c\ g =$ cosin. $y\ (R - r) + r$. Therefore cosin. $x\ (R - r) + r + D =$ cosin. $y\ (R - r) + r$, and reducing, cosin. $y = \frac{\text{cosin. } x\,(R - r) + D}{(R - r)}$; so that the distance $if$, or $hg$, measured rectangularly between the two tangents, being added to the nat. cosin. $x$, will give the nat. cosin. of the angle $elf$, to be turned on the smaller curve. The angle $y$, subtracted from the angle $x$, gives of course the angle $bce$, to be advanced on the larger curve; or, dividing this angle by the degree of curvature of $ab$, we find the distance from $b$ to $e$ the P. C. C. proper.

If $ef$ be the second curve located, and the tangent to be touched lies *within*, it is evident that we must *retreat* upon the large curve, and, by *subtracting* D from the cosine of the angle $y$, we obtain the cosine of the angle $x$.

*Example.*—Suppose $ab$ a 3° curve located, and compounding, at $b$, into a 6° curve, which latter is continued to the right through an angle of 42°. At the P. T. we discover that the proper tangent is 64 feet to the left. We must throw our curve *out*, then—we must *advance* on the 3° curve a certain distance. How to find this distance: The radius of a 3° curve = 1910; the radius of a 6° curve = 955·4; R — $r$, therefore, = 954·6. The nat. cosin. 42° = ·7431. Now, by the formula just obtained, we have $\frac{\cos. x\,(R - r) + D}{(R - r)} = \frac{(\cdot 7431 \times 954\cdot 6) + 64}{954\cdot 6} =$ ·8101 = nat. cosin. 35° 53′. Subtracting this from 42°, we have 6° 07′, the angle to be advanced on the 3° curve; or, reducing minutes to hundredths, and dividing by 3°, we find 204 feet, the distance from $b$ to the correct P. C. C.

## ARTICLE X.

### SHOULD THE SECOND CURVE BE ONE OF LONGER RADIUS THAN THE FIRST,

OUR illustration takes simpler form, and the application of D varies *vicê versa.*

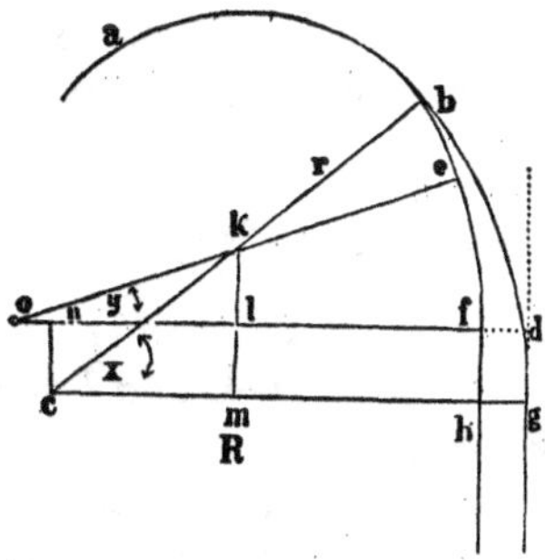


See figure, analogous to that of the previous problem. Here $o f$, $c g$ are equal and parallel radii; $f h$ a perpendicular connecting them. Draw its fellow, $c n$. Then, $n f = c h$, and, consequently, $h g = o n$. Again, letting $k m$ fall, perpendicular to $c g$, we have $c m = n l$, and $o l = c m + o n$; i. e., cosin. $y$ (R — $r$) = cosin. $x$ (R — $r$) + D.

We observe that, with a curve of this nature, in order to throw the line *farther out*, it is necessary to *go back*, toward $b$; or, having located to $g$, if the object tangent lie *within*, we must *advance* toward $e$.

*Example.*—Suppose $a\,b$ a 5° curve, $b$ the P. C. C., and $b\,g$ a 2° curve. Setting the instrument at $g$, the P. T., and turning into tangent, we find that we are a distance $h\,g$, = 53 feet, too far to the *left*. The first question is, what angle have we turned on the second curve. Let it be 28°. Now we know, that, in order to strike farther to the *right*, we must advance on the 5° curve. Consequently, D must be *added* to the cosine of 28°, to give us the cosine of the proper angle for the 2° curve; and the difference between 28° and this newly found angle will be the angle we are to advance on the 5° curve. Thus:—the rad. of a 5° curve = 1146 feet, that of a 2° curve = 2865 feet, and their difference = 1719 feet. The nat. cos. of 28° = ·8829. Then $\frac{(\cdot 8829 \times 1719) + 53}{1719} = \cdot 9137$, = nat. cos. 23° 58'. This, subtracted from 28°, leaves 4° 02', = 80 feet, from *b* to the correct P. C. C.

*Synopsis of the preceding formulæ.*

Call D the distance between tangents as before, *a* the angle of the second curve located, and *b* the angle of the same curve to be substituted for it.

First, when the second curve has the *smaller* radius—

Tangent falling *within* the point, cosine $b = \frac{\cos. a\,(R - r) + D}{(R - r)}$.

Tangent falling *without* the point, cosine $b = \frac{\cos. a\,(R - r) - D}{(R - r)}$.

Second, when the second curve has the *larger* radius—

Tangent falling *within* the point, cosine $b = \frac{\cos. a\,(R - r) - D}{(R - r)}$.

Tangent falling *without* the point, cosine $b = \frac{\cos. a\,(R - r) + D}{(R - r)}$.

Very little attention will familiarize these formulæ, and render the field practice easy.

# ARTICLE XI.

**HAVING LOCATED THE COMPOUND CURVE *a b d*, TERMINATING IN THE TANGENT *d f*, IT IS REQUIRED TO FIND THE P. C. C. *b*, AT WHICH TO COMMENCE ANOTHER CURVE OF GIVEN RADIUS, WHICH SHALL ALSO TERMINATE TANGENTIALLY IN *d f*.**

PLOT the curves as per figure. From *c* let fall *c g*, perpendicular to the tangent, *d f*. From *k* and *i*, the lesser centres, drop *k m*, *i l*, perpendicular to *c g*. Call the great radius R, the smaller radius *r*, and the *intended* radius of

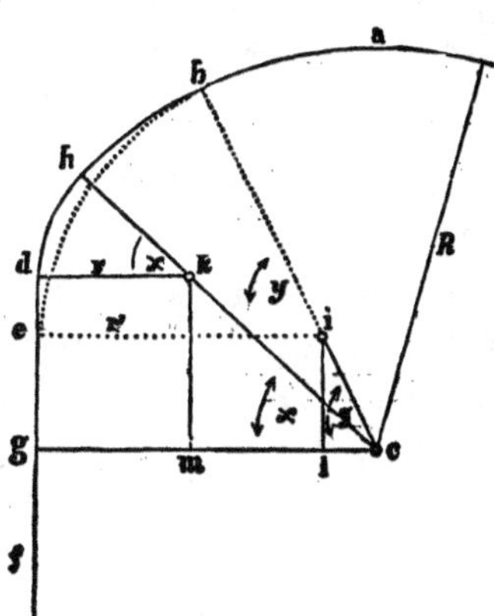


the second curve *r'*. Likewise name *h k d*, the angle of the small curve located, *x*, and *b i e* the angle to be found for the *proposed* curve, *y*. Now, the tangent *d f*, and the curve *a b*, lying unstirred, the line *c g* is an unvarying distance, and it is made up of the lines *c m* + *m g*, i. e., *c g* = (R — *r*) cosin. *x* + *r*. It also consists of the lines *c l*

$+ lg$, i. e., $cg = (R - r') \text{ cosin. } y + r'$, and, reducing, nat. cosin. $y = \frac{(R - r) \text{ cosin. } x + r - r'}{(R - r')}$. This, therefore, is the formula by means of which we can ascertain the point $b$, as follows:—

*Example.*—Imagine a 2° curve, *a h*, compounding into a 6° curve, *h d*, which terminates at *d*, in the tangent *d f*. The tangent lies well; the curve *a h* likewise; but it is desired to throw the line to the left, on better ground, between *d* and *h*, by means of an intercalary 4° curve. We wish, then, to know the distance, *h b*, back to the new P. C. C.

The radius of a 2° curve = 2865 feet, of a 6° curve = 955·4 feet, and their difference $(R - r) = 1909{\cdot}6$. The radius of a 4° curve = 1433, and the difference $(R - r')$ is, therefore, 1432 feet. Let *h k d*, the angle turned on the 6° curve, be 41°, the nat. cos. of which = ·7547.

Then, $\frac{(\cdot 7547 \times 1909{\cdot}6) + 955{\cdot}4 - 1433}{1432} = \cdot 6728$, = nat. cosin. 47° 43′. Subtracting 41°, we have 6° 43′, the angle *h c b*. Reducing minutes to hundredths, and dividing by 2°, we find 336 feet to be the distance from *h* to *b*. A 4° curve of 47° 43′, traced from this latter point, will terminate in the tangent *e f*.

## ARTICLE XII.

### IF THE LATTER CURVES HAVE LARGER RADII THAN THE FIRST,

THE solution retains its shape and simplicity.

Draw the figure as above, and, for the sake of uniformity, name the radii as before. The curve $a\,b$, and tangent $d\,f$, being constant, the distance $m\,g$, or $d\,n$, is here constant. Call it A. Now A, in the first place, is equal to $d\,i - n\,i$, i. e., $= r - (r - \mathrm{R})$ cosin. $x$; and, in the second place,

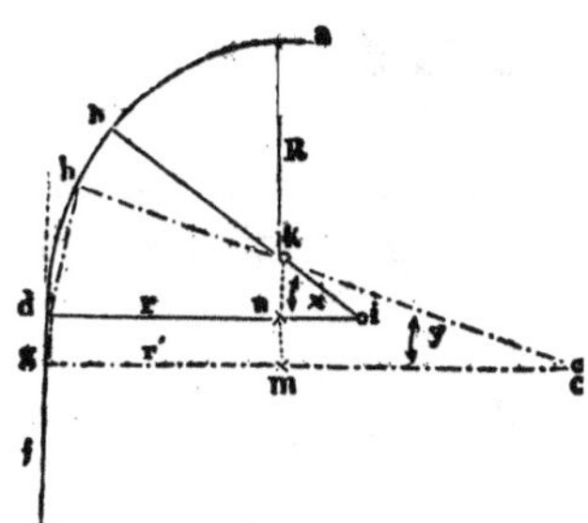


it is equal to $g\,c - m\,c$, $= r' - (r' - \mathrm{R})$ cosin. $y$; wherefore $r - (r - \mathrm{R})$ cosin. $x = r' - (r' - \mathrm{R})$ cosin. $y$, and consequently cosin. $y = \dfrac{(r - \mathrm{R}) \text{ cosin. } x + r' - r}{(r' - \mathrm{R})}$.

*Example.*—Suppose $a\,b$ a 7° curve, compounding, at $b$, into a 5° curve, $b\,d$, which latter subtends an angle of 38°, and terminates in the tangent $d\,f$. We wish to substitute a terminal 2° curve, $h\,g$, and to know the position, $h$, of the new P. C C.

The radius of a 7° curve = 819 feet = R. $r$ and $r'$,

the radii respectively of 5° and 2° curves, are equal to 1146, and 2865 feet. $r$ — R, therefore, = 327, and $r'$ — R = 2046 feet. The nat. cosin. of 38° = ·788. Then, by the formula, $\frac{(\cdot 788 \times 327) + 2865 - 1146}{2046} =$ ·9661, = the nat. cosin. of 14° 57′, the angle to be turned on the 2° curve. Subtracting this from 38°, we have 23° 03′, the angle to be continued on the 7° curve. Reducing minutes to hundredths, and dividing by the degree of curvature, 7°, we find 329 feet, the distance from $b$ to the new P. C. C. $h$.

---

## ARTICLE XIII.

### TO CHANGE A P. R. C., SO THAT THE SECOND CURVE SHALL TERMINATE IN A TANGENT PARALLEL TO A GIVEN TANGENT.

Let $a\,d\,l$ be the reverse curve, located and terminating

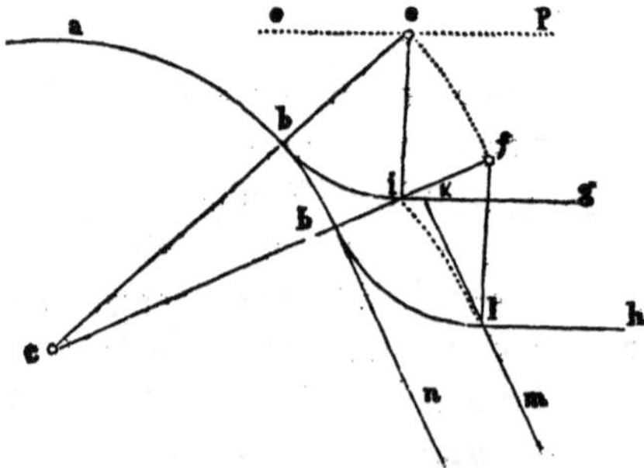

in the tangent $l\,h$. Call the radius $c\,b$, R, and the radius $b\,e$, $r$. Suppose $i\,g$ the given tangent. At a distance from it equal to *i e*, the radius of the second curve, draw the parallel line, *o p*. With *c* as a centre, and radius *c f*, $= R + r$, describe the integral curve, *f e*, cutting *o p* in *e*. *e*, then, is the centre of the curve adjusted.

## *Application.*

Place the transit at the P. T., *l*, and turn into a tangent, *l m*, parallel to *d n*, the common tangent of the two curves at *d*. Unless some wide mistake has been made, the distance *l k*, measured along this line to *i g*, the tangent proper, will be about equal to the distance *e f*, and we shall have the proportion, $cf : fe :: cd : db$, i. e., $R + r : ef :: R : db$, which gives $\frac{fe \times R}{R + r}$, as a simple formula for finding the distance back from *d* to *b*, the correct P. R. C. This rule, though sufficiently true for most cases, is not mathematically justifiable. It will be seen that *e f*, or its equal *i l*, the distance we wish to measure, is a *curving* distance, part of the circumference of a circle concentric with *a b*. Its radius is $(R + r)$, therefore its degree of curvature $= \frac{5730}{(R + r)}$, or, more simply, equals the *product* of the degrees of curvature of the curves composing the reverse, divided by their *sum*. To be strictly accurate, then, set the instrument at *l*, turn into tangent *l m* as before, and trace the curve *i l*, until it strikes the tangent *i g*. The angle which *i l* subtends, being divided by the degree of curvature of *a b*, will give the distance, *d b*, to the P. R. C. proper. The curve retraced from *b*, will terminate tangentially in *i g*, and its angle, *b e i*, will be equal to *d f l* — *d c b*.

*Example.*—Let *a d l* be a reverse curve, composed of a 3° curve, *a d*, and a 6° curve, *d l*. Let the angle *d f l* be equal to 52°, and suppose the distance *l f* to have been found 34 feet. Being part of a 2° curve, it therefore subtends a central angle of 41′. This corresponds to a distance of 23 feet, to be gone back on the 3° curve, and 52°·00 — 41′ = 51° 19′, the angle to be turned from *h* on the 6° curve, in order to strike the tangent *i g*.

---

## ARTICLE XIV.

### HOW TO PROCEED WHEN THE P. C. IS INACCESSIBLE.

In the figure, drawn to illustrate this case, let *e* be the point of curvature, *c a* the tangent, and *c k e* the curve. Now the angle *d c e*, included between the tangent and

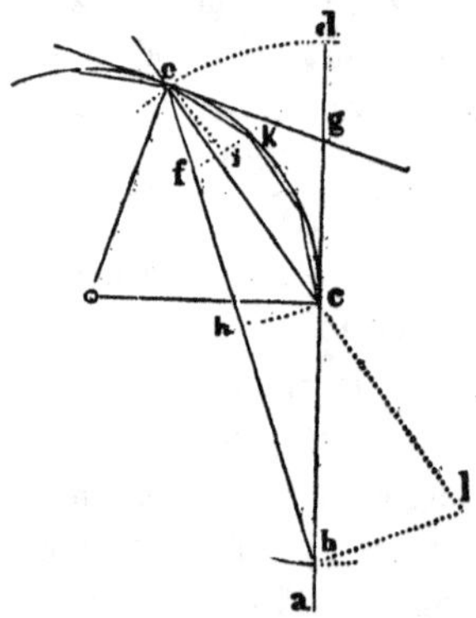


any chord, as *c e*, fixing the point *e*, is known. Make *c b* along tangent, equal to *c e*, and connect *b e*. If a circle were now described from *c* as a centre, with radius *c e* or *c b*, *d*, *e*, and *b*, would be points in its circumference, and the angle $d\,b\,e$ at once proven equal to half the angle $d\,c\,e$. With proof precisely similar, $d\,c\,e =$ half of $d\,g\,e$, and, consequently, $d\,b\,e$ is equal to one-fourth of the central angle subtended by the chord $c\,e$.

*Example.*—Suppose $c$ to be the inaccessible point of curvature of a 6° curve, $c\,k\,e$. It is concluded to run to the third station, $e$. First we must calculate the length of the chord $c\,e$. The angle $d\,c\,e = 9°$, and from Art. VII. we have

Rad. of 1 : 955·4 :: nat. sin. 9° $=$ ·1564 : $\frac{e\,c}{2}$, whereby $e\,c$ is shown equal to 298·8. Place the transit then at $b$, 298·8 feet distant from the P. C., and deflect to the left an angle of 4° 30′, equal to half the angle $d\,c\,e$. This is in line to $e$, and $b\,e$ must likewise be calculated as follows:

In the triangle $b\,c\,h$ we have

Rad. of 1 : nat. cosin. 4° 30′ $=$ ·9969 :: $b\,c =$ 298·8 : $b\,h = \frac{b\,e}{2}$, whereby $b\,e$ is shown equal to 595·7 feet. Arriving at $e$, the index reads 4° 30′. Sight back to $b$, turn to 18°, and the telescope will be in tangent. Suppose, however, that having reached $f$, 100 feet from $e$, this latter point is also found inaccessible. We find $k$ a different point in the curve, thus:—The angle $f\,e\,g$ = 18° — 4° 30′ = 13° 30′, and the tangential angle $g\,e\,k$ = 3°. Consequently the angle $f\,e\,k$ = 10° 30′, and, drawing the bisecting line $e\,i$, we have, in the triangle $e\,f\,i$,

Rad. of 1 : nat. sin. 5° 15′ = ·0915 :: $e\,f$ = 100 : $f\,i$ = 9·159 feet. Therefore $f\,k$ = 18·318 feet, and the angle $e\,f\,k$ = 90° — 5° 15′ = 84° 45′. At $f$ deflect this angle to the right, and measure the distance $f\,k$ carefully with the rod. At $k$, sighting back to $f$, and turning the equal angle $f\,k\,e$, the telescope will be directed to $e$, and the curve may be continued.

If it is inconvenient to run the line $b\,e$, the point $e$ may be reached thus:—Fix the P. C. Find the tangential distance *d e*, corresponding to the angle *d c e*. Carefully with the rod lay off *b l*, equal to it, at right angles to *b e* Set the transit at *l*, and, in line with *c*, put in *e*.

The distance *b l* should not exceed 10 or 12 feet.

Note.—The foregoing illustrations will apply when the P. T. is likewise inaccessible.

---

## ARTICLE XV.

### TO AVOID OBSTACLES IN THE LINE OF CURVE.

Let *b k h* be the curve. We can either follow the tangents *h d*, *d b*, or trace a parallel curve, *g a*, *within* the first; which tracing is effected thus:—Set the instrument

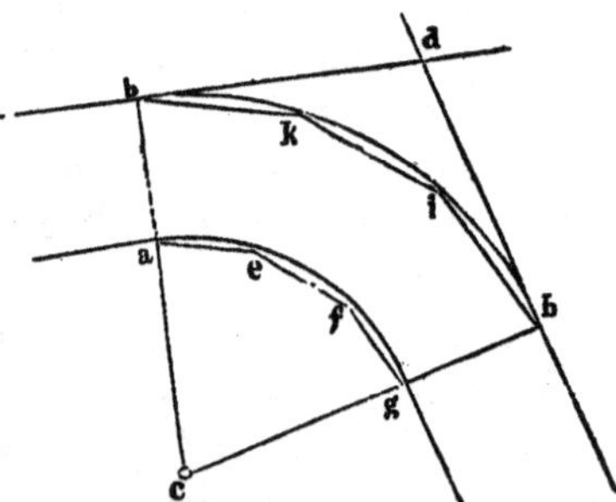


at *h*, the P. C., and offset any distance *h g*, at right angles to the tangent *h d*. It will be observed that as the distance *h g increases*, the distance *g a decreases*, whilst the angle subtended by *g a* remains equal to that subtended by $h\,b$; i. e., our *deflexions* on the offset curve stand unchanged, but the corresponding *chords*, $g\,f$, $f\,e$, &c., are *less* than their equivalents, $h\,i$, $i\,k$, &c., along $h\,b$. To find their length, $h\,i$, $i\,k$, &c., being equal to 100 feet, we have the proportion, $c\,h : c\,g :: h\,i : x$; i. e., R : rad. — $h\,g$ : : 100 feet : $x$, where $x$ symbols the unknown chord. Now set the transit at $g$, turn into tangent parallel to $h\,d$, and with the shortened chord, fix $f\,e\,a$. Rectangularly to the tangents at these points, and distant $h\,g$, will be $i$, $k$, $b$ of the curve proper.

*Example.*—Let $b\,h$ be a 4° curve, and the offset distance 85 feet. The radius then is 1433, and

1433 : 1348 : : 100 : 94, the short chord.

*To follow the tangents*, suppose the angle $b\,c\,h$ = 42°. Then by Art. V. we find the tangent $h\,d$ = 550 feet, which distance we duly measure, and, at $d$, deflecting 42°, lay off an equal distance to $b$, the point of tangency.

# ARTICLE XVI.

HAVING GIVEN THE ANGLES $d\,b\,k$, $m\,k\,l$, AND THE DISTANCE $b\,k$, IT IS REQUIRED TO FIND THE RADII $c\,e$, $e\,f$ OF THE EASIEST REVERSE CURVE WHICH SHALL UNITE $a\,d$, $k\,m$.

THE angle $d\,b\,e$ is equal to the angle $a\,c\,e$, half of which is $b\,c\,e$. So likewise $e\,f\,k$ is equal to half of $l\,k\,m$.

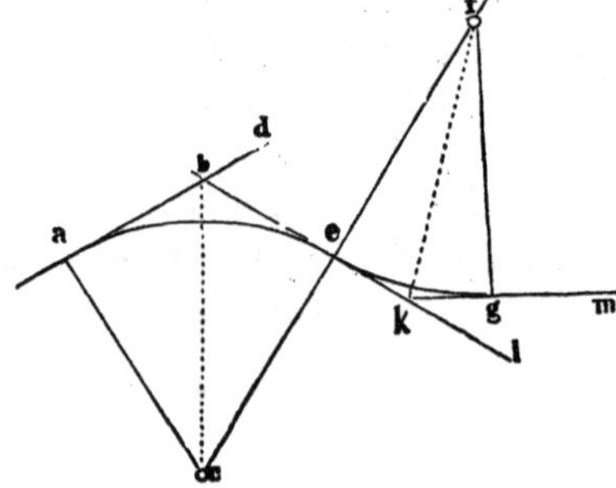


Then, [nat. tang. $b\,c\,e$ + nat. tang. $e\,f\,k$] : nat. tang. $b\,c\,e$ :: $b\,k$ : $b\,e$, and $b\,k - b\,e = e\,k$. Wherefore rad. $c\,e$ $= \dfrac{b\,e}{\text{nat. tang. } b\,c\,e}$ and rad. $e\,f = \dfrac{e\,k}{\text{nat. tang. } k\,f\,e}$.

*Example.*—Suppose the angle $d\,b\,e = 54°\ 30'$, the angle $l\,k\,g = 33°\ 20'$, and the distance $b\,k = 832$ feet.

Therefore the angle $b\,c\,e = 27°\ 15'$, the nat. tang. of which is ·5150, and the angle $e\,f\,k = 16°\ 40'$, the nat. tang. of which is ·2994. The sum of the tangents = ·8144. Then, to find $b\,e$, we have

As ·8144 : ·5150 :: 832 : 526, and subtracting this from $b\,k$, we have $e\,k = 306$ feet.

Again, the radius $c\,e = \dfrac{526}{\cdot 515}$, and the radius $e\,f = \dfrac{306}{\cdot 2994}$, = 1022 feet.

## ARTICLE XVII.

HAVING GIVEN THE CURVE $fg$, LOCATED AND TERMINATING IN THE TANGENT $g\,m$, IT IS REQUIRED TO FIND WHERE A CURVE OF DIFFERENT RADIUS WILL TERMINATE IN A TANGENT PARALLEL TO $g\,m$.

LET $fg$ be the curve located, and $fh$ the curve proposed, commencing at the common point $f$, and terminating in the parallel tangents, $g\,m$, $h\,l$. We wish to find the length and direction of the line $g\,h$, connecting the points of tangency.

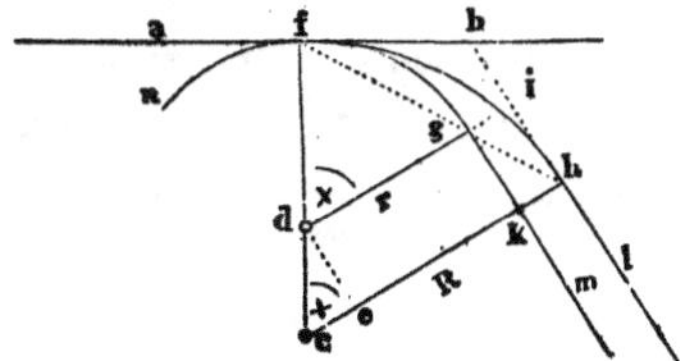

Call the radii respectively R and $r$, and the central angle, $f\,d\,g$ or $f\,c\,h$, $x$. Now $g\,k = d\,e$, = sine of $x$, $c\,d$ being radius; i. e., = nat. sin. $x \times (R - r)$. Again, the angle $k\,g\,h = f\,h\,b = \frac{x}{2}$, and $g\,k$ = cosin. $k\,g\,h$, $g\,h$ being radius, i. e., $g\,k$ = nat. cos. $\frac{x}{2} \times g\,h$, and, consequently, nat. cos. $\frac{x}{2} \times g\,h$ = nat. sin. $x \times (R - r)$, wherefore

$$g\,h = \frac{(R - r)\ \text{nat. sin. } x}{\text{nat. cos. } \frac{x}{2}}.$$

It will be observed that the angle $k\,g\,h$, included between the tangent to a curve at any point $g$, and the line $g\,h$ connecting $g$ with an equivalent point $h$ in any other curve $f\,h$ commencing at the same P. C., $f$, and turning in the same direction, is invariably equal to half the common central angle, $f\,d\,g$ or $f\,c\,h$.

*Example.*—Let $f\,g$ be a 7° curve, subtending a central angle, $f\,d\,g$, of 44° 26′. Having arrived at $g$, the P. T., and turned into tangent, $g\,m$, it is desired to fix the P. T., $h$, of a 4° curve which shall likewise subtend an angle of 44° 26′. Here the radii are, respectively, 819 and 1433 feet, and $R - r = 614$. The nat. sine of 44° 26′ = ·700, and the nat. cosine of half this angle, viz.: 22° 13′ = ·9258. Then, by the formula, $g\,h = \frac{614 \times \cdot 7}{\cdot 9258} = 464{\cdot}2$. Deflecting, therefore, to the left, an angle of 22° 13′, and laying off the distance 464·2 feet, we arrive at the point $h$. Move to $h$, sight back to $g$, and a deflexion of 22° 13′ to the right will direct the telescope along the tangent $h\,l$.

If $h$ were the P. T. located, and $g$ the point required, the same angle and distance would apply.

## ARTICLE XVIII.

HAVING THE CURVE *nfg* LOCATED, AND TERMINATING IN THE TANGENT *g m*, IT IS REQUIRED TO FIND THE POINT *f*, WHEREAT TO COMPOUND WITH ANOTHER CURVE OF GIVEN RADIUS, WHICH SHALL TERMINATE IN *i l*, PARALLEL TO *g m*.

[See previous figure.]

NAME the radii and angle as before. Measure the distance *g i*, between the tangents, and call it D. Then *c h* is equal to *c e* + *e k* + *g i* or *k h*; i. e., $R = (R - r)$ cosin. $x + r + D$, or, transposing, cosin. $x = \frac{R-(r-D)}{(R-r)}$ Thus discovering the angle *f d g*, divide it by the curvature *nfg*, and we have the distance, *gf*, to the P. C. C. *f*. The second curve, traced from this point, will terminate tangentially in *h l*.

*Example.*—Let the curvatures equal those of the last problem, and suppose the distance *g i* to be 175·6 feet. Then, by the formula, cosin. $x = \frac{1433 - 994{\cdot}6}{614} = {\cdot}7143$ = cosin. 44° 26′. Reducing minutes to hundredths, and dividing by 7°, we have 635 feet, the distance back to the P. C. C.

## ARTICLE XIX.

HAVING GIVEN A TANGENT *a b*, AND A CURVE *b k*, LOCATED, IT IS REQUIRED TO FIND THE POINT *d*, OR *f*, AT WHICH TO COMMENCE A CURVE OF GIVEN RADIUS, WHICH SHALL BE TANGENT TO BOTH.

DRAW the radius *b c*, and call it R. Name the radius of the other curve *r*. Make *b e* equal to *r*, and, through *e*, draw the line *e l*, parallel to the tangent *a b*. From *c* as a centre, with *c g*, $=(\text{R}+r)$, as radius, sweep the arc of a circle; which arc will intersect *e l* at *g*, and, from the equidistances, prove *g* the centre of the other curve touching *a b*, *b k*, tangentially, at the points *d* and *f*.

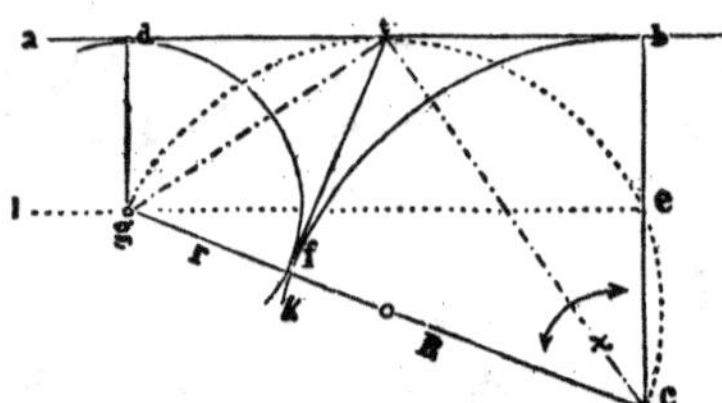


Now, to find the point *f* for purposes of location, we must know the angle *b c f*. Call it $x$. In the triangle *c g e*, we have *c g* : *c e* : : radius : cosin. *g c e*, or *b c f*; i. e., $(\text{R}+r):(\text{R}-r)::1:$ nat. cosin $x$, wherefore nat. cosin. $x=\frac{(\text{R}-r)}{(\text{R}+r)}$. So that, having divided the *difference* of the radii by their *sum*, we shall find opposite to the quotient, in the table of nat. cosines, the angle *b c f* required. This angle, divided by the degree of curvature of *b k*, will give the distance from $b$ to the P. R. C., $f$; and $180° - x$ will equal the angle $f\,g\,d$, to be turned on the other curve.

*Another plan*, which may be preferred, for finding the point $d$, is as follows:

Make $c\,g$ the diameter of a semicircle, $c\,t\,g$. This semicircle is tangent to $b\,d$ at $t$, and $t\,f$, perpendicular to $c\,g$, is a common tangent to the curves $b\,k$, $d\,f$. We have then, $c\,f \times f\,g = t\,f^2$, i. e., $R \times r = \text{tang.}^2 \frac{x}{2}$. Multiplying the radii together, and extracting the square root of the product, we find the distance $t\,f$, or its equal $t\,b$, which, *doubled*, is the distance $b\,d$, from $b$ to the point of curvature, $d$.

*Example.*—Suppose $k\,b$ a 3° curve, tangent to the line $a\,b$. We wish to know the point $f$, at which to begin a 5° curve, which shall be tangent to both. Here $R = 1910$, and $r = 1146$, wherefore $(R + r) = 3056$, $(R - r) = 764$, and $\frac{(R - r)}{(R + r)} = \frac{764}{3056}$, $= \cdot 25$, $=$ nat. cosin. 75° 31′. Reducing minutes to hundredths, and dividing by 3°, we have 2517 feet, the distance from $b$ to the P. R. C., $f$.

If the point $d$ be required, we have $\sqrt{1910 \times 1146} = 1478 \cdot 9$. Doubling this result we find $2957 \cdot 8$ feet, the distance from $b$ to the P. C., $d$.

If the radii are equal, of course the distance $b\,d$ is equal to their *sum*, and the angle $x$ is a right angle.

## ARTICLE XX.

### TO RUN A TANGENT TO TWO CURVES.

LET $g\,b$, $e\,k$, be the two curves. First plot them carefully to a large scale, and, finding from this plot an approximate P. T., say $e$, run the tangent $ef$. Now the chord in the curve $g\,b$, to which $ef$ is parallel, is known, and consequently the shortest distance, $fg$, between the tangent and the curve may be measured. Then $ef : fg$ :: rad. : tang. $feg$, and nat. tang. $feg = \frac{fg}{ef}$. *Practically*, this angle $feg$ will be found equal to $g\,a\,b$ or $d\,c\,e$, so that, dividing it by the degree of curvature of $k\,e$, we shall have the distance, $e\,d$, to the proper P. T.

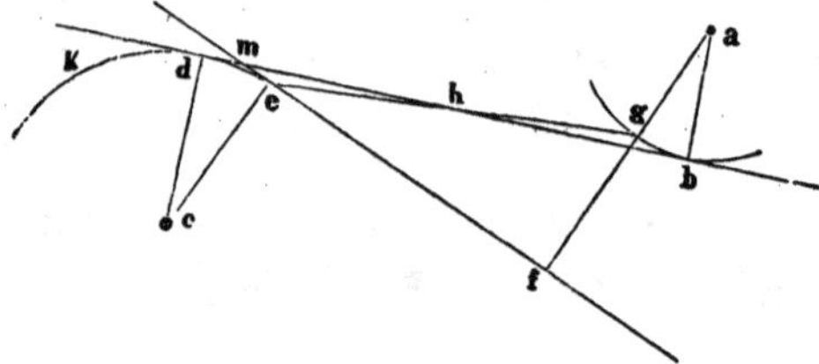


Strictly speaking, the angle $gef$ is too great by the small angle $b\,h\,g$. There are two modes of calculating this latter, but they are both complex, and in any but a most unusual case, the above rule is sufficiently accurate.

*Example.*—Let $k\,e$ be a 5° curve. Suppose the tangent $ef = 1632$ feet, and the distance $gf = 42$ feet,

Then $\frac{42}{1632} = \cdot 0257 =$ nat. tang. of 1° 28′. Reducing minutes to hundredths, and dividing by 5°, we have 29 feet, the distance back to the correct P. T.

Should the curves turn in the same direction, or should $e$ be a fixed point from which to run a tangent to $g\,b$, the above illustration is still applicable.

---

## ARTICLE XXI.

### ORDINATES.

#### TO FIND THE MIDDLE ORDINATE TO ANY GIVEN CHORD, IN A CURVE OF ANY GIVEN RADIUS.

1st. (See figure in Art. XVI.) $k\,c = \sqrt{a\,c^2 - a\,k^2}$, and $a\,c$ or $b\,c - k\,c =$ the ordinate required. That is, from the square of the radius subtract the square of half the chord, and take the square root of the remainder from radius, for the middle ordinate.

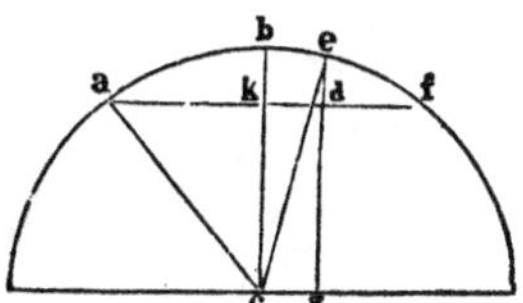


*Example.*—The radius $a\,c$ being 819 feet, and the chord $a\,f$ 100 feet, to find the middle ordinate, $b\,k$.

Here $a\,c^2 - a\,k^2 = 670761 - 2500, = 668261$, the square root of which is $c\,k, = 817{\cdot}5$, which taken from radius 819, leaves 1·5, the required middle ordinate.

2d. Subtract the nat. cosine of the tangential angle from 1, and multiply the remainder by radius.

*Example.*—Suppose $a\,b$ a 7° curve. Here the nat. cosine of 3° 30′, the tangential angle, is ·9981, which, subtracted from 1, leaves ·0019. Multiplying this latter by radius 819, we have 1·5, the middle ordinate as before.

### HAVING GIVEN THE MIDDLE ORDINATE, TO FIND ANY OTHER.

1st. $e\,g = \sqrt{c\,e^2 - c\,g^2}$, and $e\,d = e\,g - g\,d$ or $c\,k$.

*Example.*—Suppose the distance $k\,d$ or $c\,g$ to be 20 feet. Thence $c\,e^2 - c\,g^2 = 670761 - 400 = 670361$, the square root of which is $e\,g$, = 818·76. Taking from this the distance $g\,d$ = 817·5, we have the ordinate $e\,d$ = 1·26.

2d. Multiply the ordinates of a 1° curve by the deflexion angle of the curve whose ordinates are required. This is an approximation sufficiently exact for railway curves.

## ARTICLE XXII.

### TO FIND THE RADIUS CORRESPONDING TO ANY DEFLEXION ANGLE, AND TO EQUAL CHORDS OF ANY GIVEN LENGTH.

HERE the deflexion angle, *d b a*, is of course equal to the central angle *d c b*, subtended by the given chord *d b*, and the triangle *c d b* being isosceles we have $\frac{180° - d\,c\,b}{2}$ $= c\,d\,b$, or *d b c*.

Then nat. sin. *d c b* : nat. sin. *c b d* :: *d b* : *c d*, the radius.

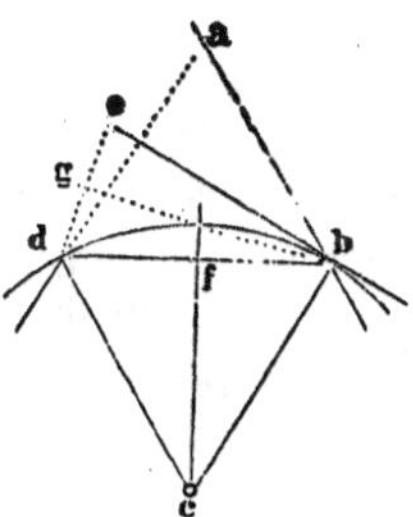


*Example.*—Let the deflexion angle be 5°, and the chord 100 feet. Required the radius. The nat. sine of 5° = ·0872 and $\frac{180° - 5°}{2} = 87° \; 30'$, the nat. sine of which is ·9990.

Then ·0872 : ·9990 :: 100 : radius = 1146 feet.

*An approximate result*, sufficiently accurate for all purposes of railway location, may be found by simply dividing 5730 by the deflexion angle.

*To find the deflexion angle corresponding to any given*

*radius and chord.*—In this case, we have $c\,d : d\,f :: $ rad. of 1 : nat. sin. of half the deflexion angle; therefore, divide half of the chord by the radius of the curve; the quotient will be the nat. sine of half the deflexion angle.

*Example.*—Radius as before 1146 feet, and chord 100 feet. Then $\frac{50}{1146} = \cdot 0436$, the nat. sin. of 2° 30′. Doubling this result we have the deflexion angle, viz. 5° 00′.

*To find the deflexion distance with chord of* 100 *feet and any radius.*—Divide the constant number 10000 by the radius in feet; the quotient will be the deflexion distance:—for the deflexion distance with a radius of 10000 feet is 1 foot, and the deflexion distances for other radii increase inversely as the radii.

*Example.*—What is the deflexion distance for a 5° curve, the chord being 100 feet? Here $\frac{10000}{1146} = 8\cdot 72$ feet, the deflexion distance.

*To find the deflexion distance with any given chord and radius.*—The deflexion distance is equal to twice the natural sine of half the *deflexion* angle, multiplied by the chord. Thus, the chord being 100 feet, and the deflexion angle 5°, we find the nat. sine of 2° 30′ equal to ·0436, which doubled, and multiplied by 100, gives 8·72 feet, as above.

*The tangential distance, with any radius and chord,* is in like manner equal to twice the nat. sine of half the *tangential* angle multiplied by the chord. Thus, the tangential angle being 2° 30′ and the chord 100 feet, we find the nat. sine of 1° 15′ = ·0218. Multiplying by 2, and 100, we have 4·36 feet, the tangential distance.

For all curves under 11°, the tangential distance is equal to half the deflexion distance.

## ARTICLE XXIII.

### OF EXCAVATION AND EMBANKMENT.

A RAILWAY line having been located, the first office duties are to map it down, to make a continuous profile, and an approximate estimate of the cost of grading, &c. The "elevation" of each stake above or below a certain assumed base being fixed, the profile is drawn, on paper prepared for the purpose, to a horizontal scale of 400 feet, and a vertical scale of 40 feet, to the inch. This distorted picture presents at a glance, in compact shape, the undulations of the surface, and from it grades are adopted, to balance as nearly as possible the excavation and embankment. By these grades, noted in the record, the cutting or filling is ascertained at each station. Suppose, for instance, that the elevation of station 40 is 12 feet, and that of station 100 is 72 feet. The distance between 40 and 100 is 6000 feet, and the difference between 12 and 72 is 60 feet. Consequently, to connect those points, we require an ascending grade of 1 in 100 feet. Grade at station 54 is therefore 26 feet, and at station 60, 32 feet. If the elevation at station 60 be 38 feet, of course we have 6 feet of excavation, and this is marked in our estimate sheets + 6. If the elevation at that point should be 27 feet, 5 feet of embankment is the consequence, which is marked accordingly, — 5; *plus* indicating excavation, and *minus* embankment.

The usual slope for embankment is 1½ feet horizontal to 1 foot vertical, making an angle of about 34° with the horizon; that for earth cut, 1 to 1, or 45°, and for rock cut ¼ to 1, or 76°. The slopes of the ground surface at each station being known, together with the breadth of the intended roadway, we are prepared to calculate the cross-sectional areas, and from them to determine the cubic yards of excavation and embankment.

To facilitate these operations the following rules were prepared. With Trautwine's common diagram, and the table of squares and square roots appended to the volume, they will be found an observable assistance.

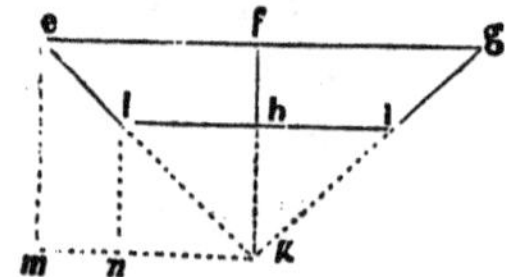


Suppose $l\,i$ the road bed, and $f\,h$ the depth of an ordinary clay cut. Produce the side slopes until they meet in the point $k$. Then the angle being 45°, $e\,f = f\,k$, and $e\,f^2 =$ twice the triangle $e\,f\,k$, $=$ the triangle $e\,k\,g$. For like reasons $l\,h = h\,k$, and $l\,h^2 =$ triangle $l\,k\,i$. But the triangle $l\,k\,i$, taken from the triangle $e\,k\,g$, leaves the area $e\,l\,i\,g$, to find which we have therefore the following

*Rule.*—To half the breadth of the roadway, add the depth of the cut. From the square of this sum, subtract the square of half the breadth of the roadway for the area.

*Example.*—Suppose the breadth of the roadway to be 32 feet, and the depth $h\,f$, 7 feet. Half of $32 = 16$, the square of which, viz., 256, becomes a constant subtrahend. $16 + 7 = 23$, and a reference to the table shows the square of 23 equal to 529. Therefore $529 - 256 = 273$, the area required in square feet.

If there be a regular slope, as $e\,g$ (p. 58), we can plot it on the diagram, and read at once the side cuts, $g\,k$, $e\,n$. Then the area is equal to that of the trapezoid $e\,n\,k\,g$, *minus* the two triangles $e\,n\,h$, $m\,k\,g$.

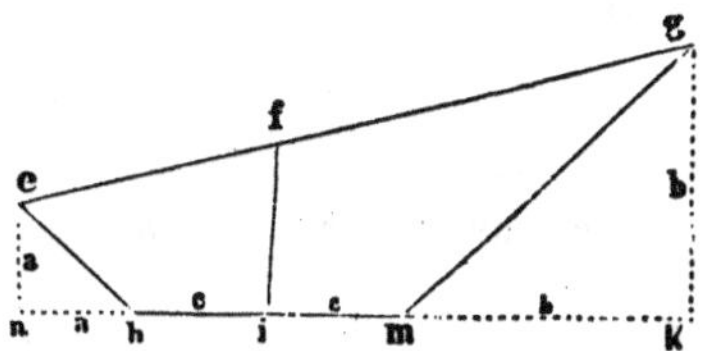


Calling $e\,n$, or its equal $n\,h$, $a$, $m\,k$ or $g\,k$, $b$, and half the breadth of the roadway, $c$, we have the triangle $m\,k\,g$ $= b \times \frac{b}{2} = \frac{b^2}{2}$. The triangle $e\,n\,h$ is in like manner $= \frac{d^2}{2}$, and the area of the trapezoid $= a + b + 2\,c \times \frac{a + b}{2}$. Then the area required $= a + b + 2\,c \times \frac{a + b}{2} - \frac{a^2 + b^2}{2}$

$$= \frac{a^2 + 2\,a\,b + 2\,a\,c + 2\,b\,c + b^2}{2} - \frac{a^2 + b^2}{2}$$

$$= \frac{2\,a\,b + 2\,a\,c + 2\,b\,c}{2} = (a + b)\,c + a.b.$$

Should the slopes be $\frac{1}{4}$ to 1, as in rock cut, the area might, in similar-wise, be shown equal to $(a + b)\,c + \frac{a.b}{4}$, and that of embankment, where the slopes are $1\frac{1}{2}$ to 1, equal to $(a + b)\,c + \frac{3\,a.b}{2}$. Wherefore, we arrive at the following general rule for finding cross-sectional areas, where the ground surface is a regular declivity:

Multiply the *sum* of the side cuttings by half the breadth of the roadway, and mark the product. Multiply the *product* of the side cuttings by the ratio of the side slopes to 1, and add this result to the previous product marked for the area.

*Example* 1.—Earth excavation—road bed 32 feet, side

slopes 1 to 1, right cutting 12 feet, left cutting 3 feet. Required the cross-sectional area. Here $(a + b)\ c + a\ b = (12 + 3)\ 16 + 12 \times 3 = 240 + 36 = 276$ square feet, the area required.

*Example* 2.—Rock cut—road bed 28 feet, side slopes $\frac{1}{4}$ to 1, cuttings as before. Required the area. Here $(a + b)\ c + \frac{a \times b}{4} = 210 + \frac{36}{4} = 219$ square feet, the area.

*Example* 3.—Embankment—roadway 27 feet, side slopes $1\frac{1}{2}$ to 1, cuttings as before. Here $(a + b)\ c + \frac{3\ (a \times b)}{2} = 202{\cdot}5 + 54 = 256{\cdot}5$ square feet, the cross-sectional area.

Other formulæ might be given for varying surface slopes, but they would involve a simple matter, and require more time for calculation than a division of the diagram area into triangles.

To find the cubic content of an excavation or embankment:—

*Multiply half the sum of the two end areas by the distance between them;* thus: supposing 282 and 310 to be the end areas, and 100 feet the distance, we have $\frac{(282 + 310)}{2} \times 100 = 29600$ cubic feet $= 1096{\cdot}3$ cubic yards. Or we may multiply the sum of the end areas by 100, and divide the product by 6 and 9, the factors of 54, which is the number of cubic feet contained in two cubic yards; thus, $\frac{59200}{6} = 9866{\cdot}6$, and this divided by $9 =$ $1096{\cdot}3$, as above.

### PRISMOIDAL FORMULA.

*To the two end areas add four times the middle area, and multiply the sum by one-sixth of the length of the prismoid.* Thus: from the foregoing example, the sum of the end areas, 592, added to four times their mean, 1184, gives 1776, and 1776 × 16·7 = 29659·2 cubic feet, = 1098·5 cubic yards. The former rule is approximate—sufficiently so for rough preliminary calculations, but where strict correctness is required, as in a final field estimate, the prismoidal formula should be used. It applies to all solid bodies with plane faces and parallel ends.

---

## ARTICLE XXIV.

### SIDE STAKING.

The object in side staking is to find the point where the surface of the ground intersects the slope of the road

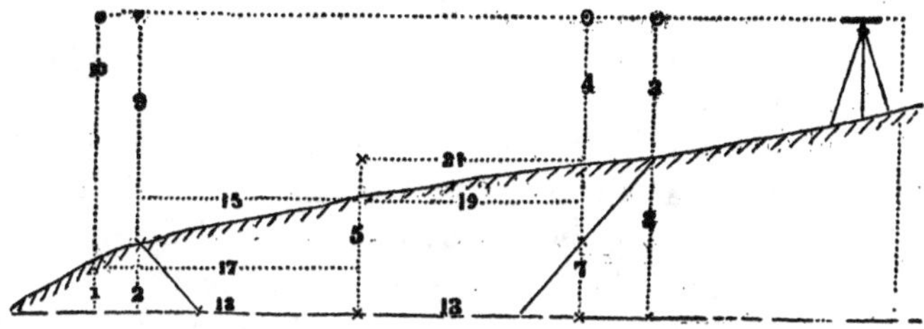


formation. For performing this work with level and rod, place the instrument in such a position as to command as many stations as possible, whether they be on the upper or lower side, and ascertain by transfers from the bench the height of your instrument with reference to grade. This will facilitate operations by giving usually small numbers to work with. Knowing your rate of grade, subtract or add, as the case may be, as you change from station to station.

Having the level in place, we will suppose on an excavation, our object is first to place the lower side stake. Let the width of the road bed be 26 feet, and the side slopes 1 to 1. Suppose the instrument to be 11 feet above grade and the centre cut 5 feet, which latter is already recorded in the field book for construction duties. Looking at the fall of the hill we judge that at the distance of 17 feet from the centre stake the descent is 1·5 feet, which would give us 3·5 feet cutting. We take an observation at that point and find that the rod reads 10 feet, which, subtracted from our height above grade, leaves us only one foot cutting, and shows that our judgment has been at fault. The distance measured is too much, for were we to increase the distance we would reduce the cutting, and it is therefore evident that we are too far from the centre. Let us then make the distance from the centre stake 15 feet, and take another sight, when the rod reads 9, showing 2 feet cutting, which, added to half the road formation, slope being 1 to 1, equals the distance measured. The stake is therefore correct. The same process applies to the upper side.

The only difference in staking out for banks is that you add 1½ times the height of the bank to the distance measured, or, as a general rule, the height of the bank multiplied by the ratio of the side slopes to unity.

Some judgment is required to be expeditious in this work, which is obtained only by experience.

6

The following is a common form of field record for construction:

| | | | | | | Left Side. | | Centre. | | Right Side. | |
|---|---|---|---|---|---|---|---|---|---|---|---|
| Sta. | Dist. | Course. | Mag. Course. | Elevation | Grade. | Dist. | A or B | A | B | Dist. | A or B |
| 530 | 100 | N 20°00 W | N 20·05 W | + 560·4 | + 555·4 | 15·0 | 2·0 A | 5·0 | | 21·0 | 8·0 A |

# NATURAL SINES AND TANGENTS.

| ′ | 0° | 1° | 2° | 3° | 4° | 5° | 6° | 7° | |
|---|---|---|---|---|---|---|---|---|---|
| 0 | 000 0000 | 017 4524 | 034 8995 | 052 3360 | 069 7565 | 087 1557 | 104 5285 | 121 8693 | 60 |
| 1 | 2909 | 7432 | 035 1902 | 6264 | 070 0467 | 4455 | 8178 | 122 1581 | 59 |
| 2 | 5816 | 018 0341 | 4809 | 9169 | 3368 | 7353 | 105 1070 | 4468 | 58 |
| 3 | 8727 | 3249 | 7716 | 053 2074 | 6270 | 088 0251 | 3963 | 7355 | 57 |
| 4 | 001 1636 | 6158 | 036 0623 | 4979 | 9171 | 3148 | 6856 | 123 0241 | 56 |
| 5 | 4544 | 9066 | 3530 | 7883 | 071 2073 | 6046 | 9748 | 3128 | 55 |
| 6 | 7453 | 019 1974 | 6437 | 054 0788 | 4974 | 8943 | 106 2641 | 6015 | 54 |
| 7 | 002 0362 | 4883 | 9344 | 3693 | 7876 | 089 1840 | 5533 | 8901 | 53 |
| 8 | 3271 | 7791 | 037 2251 | 6597 | 072 0777 | 4738 | 8425 | 124 1788 | 52 |
| 9 | 6180 | 020 0699 | 5158 | 9502 | 3678 | 7635 | 107 1318 | 4674 | 51 |
| 10 | 9089 | 3608 | 8065 | 055 2406 | 6580 | 090 0532 | 4210 | 7560 | 50 |
| 11 | 003 1998 | 6516 | 038 0971 | 5311 | 9481 | 3429 | 7102 | 125 0446 | 49 |
| 12 | 4907 | 9424 | 3878 | 8215 | 073 2382 | 6326 | 9994 | 3332 | 48 |
| 13 | 7815 | 021 2332 | 6785 | 056 1119 | 5283 | 9223 | 108 2885 | 6218 | 47 |
| 14 | 004 0724 | 5241 | 9692 | 4024 | 8184 | 091 2119 | 5777 | 9104 | 46 |
| 15 | 3633 | 8149 | 039 2598 | 6928 | 074 1085 | 5016 | 8669 | 126 1990 | 45 |
| 16 | 6542 | 022 1057 | 5505 | 9832 | 3986 | 7913 | 109 1560 | 4875 | 44 |
| 17 | 9451 | 3965 | 8411 | 057 2736 | 6887 | 092 0809 | 4452 | 7761 | 43 |
| 18 | 005 2360 | 6873 | 040 0318 | 5640 | 9787 | 3706 | 7343 | 127 0646 | 42 |
| 19 | 5268 | 9781 | 4224 | 8544 | 075 2688 | 6602 | 110 0234 | 3531 | 41 |
| 20 | 8177 | 023 2690 | 7131 | 058 1448 | 5589 | 9499 | 3126 | 6416 | 40 |
| 21 | 006 1086 | 5598 | 041 0037 | 4352 | 8489 | 093 2395 | 6017 | 9302 | 39 |
| 22 | 3995 | 8506 | 2944 | 7256 | 076 1390 | 5291 | 8908 | 128 2186 | 38 |
| 23 | 6904 | 024 1414 | 5850 | 059 0160 | 4290 | 8187 | 111 1799 | 5071 | 37 |
| 24 | 9813 | 4322 | 8757 | 3064 | 7190 | 094 1083 | 4689 | 7956 | 36 |
| 25 | 007 2721 | 7230 | 042 1663 | 5967 | 077 0091 | 3979 | 7580 | 129 0841 | 35 |
| 26 | 5630 | 025 0138 | 4569 | 8871 | 2991 | 6875 | 112 0471 | 3725 | 34 |
| 27 | 8539 | 3046 | 7475 | 060 1775 | 5891 | 9771 | 3361 | 6609 | 33 |
| 28 | 008 1448 | 5954 | 043 0382 | 4678 | 8791 | 095 2666 | 6252 | 9494 | 32 |
| 29 | 4357 | 8862 | 3288 | 7582 | 078 1691 | 5562 | 9142 | 130 2378 | 31 |
| 30 | 7265 | 026 1769 | 6194 | 061 0485 | 4591 | 8458 | 113 2032 | 5262 | 30 |
| 31 | 009 0174 | 4677 | 9100 | 3389 | 7491 | 096 1353 | 4922 | 8146 | 29 |
| 32 | 3083 | 7585 | 044 2006 | 6292 | 079 0391 | 4248 | 7812 | 131 1030 | 28 |
| 33 | 5992 | 027 0493 | 4912 | 9196 | 3290 | 7144 | 114 0702 | 3913 | 27 |
| 34 | 8900 | 3401 | 7818 | 062 2099 | 6190 | 097 0039 | 3592 | 6797 | 26 |
| 35 | 010 1809 | 6309 | 045 0724 | 5002 | 9090 | 2934 | 6482 | 9681 | 25 |
| 36 | 4718 | 9216 | 3630 | 7905 | 080 1989 | 5829 | 9372 | 132 2564 | 24 |
| 37 | 7627 | 028 2124 | 6536 | 063 0808 | 4889 | 8724 | 115 2261 | 5447 | 23 |
| 38 | 011 0535 | 5032 | 9442 | 3711 | 7788 | 098 1619 | 5151 | 8330 | 22 |
| 39 | 3444 | 7940 | 046 2347 | 6614 | 081 0687 | 4514 | 8040 | 133 1213 | 21 |
| 40 | 6353 | 029 0847 | 5253 | 9517 | 3587 | 7408 | 116 0929 | 4096 | 20 |
| 41 | 9261 | 3755 | 8159 | 064 2420 | 6486 | 099 0303 | 3818 | 6979 | 19 |
| 42 | 012 2170 | 6662 | 047 1065 | 5323 | 9385 | 3197 | 6707 | 9862 | 18 |
| 43 | 5079 | 9570 | 3970 | 8226 | 082 2284 | 6092 | 9596 | 134 2744 | 17 |
| 44 | 7987 | 030 2478 | 6876 | 065 1129 | 5183 | 8986 | 117 2485 | 5627 | 16 |
| 45 | 013 0896 | 5385 | 9781 | 4031 | 8082 | 100 1881 | 5374 | 8509 | 15 |
| 46 | 3805 | 8293 | 048 2687 | 6934 | 083 0981 | 4775 | 8263 | 135 1392 | 14 |
| 47 | 6713 | 031 1200 | 5592 | 9836 | 3880 | 7669 | 118 1151 | 4274 | 13 |
| 48 | 9622 | 4108 | 8498 | 066 2739 | 6778 | 101 0563 | 4040 | 7156 | 12 |
| 49 | 014 2530 | 7015 | 049 1403 | 5641 | 9677 | 3457 | 6928 | 136 0038 | 11 |
| 50 | 5439 | 9922 | 4308 | 8544 | 084 2576 | 6351 | 9816 | 2919 | 10 |
| 51 | 8348 | 032 2830 | 7214 | 067 1446 | 5474 | 9245 | 119 2704 | 5801 | 9 |
| 52 | 015 1256 | 5737 | 050 0119 | 4349 | 8373 | 102 2138 | 5593 | 8683 | 8 |
| 53 | 4165 | 8644 | 3024 | 7251 | 085 1271 | 5032 | 8481 | 137 1564 | 7 |
| 54 | 7073 | 033 1552 | 5929 | 068 0153 | 4169 | 7925 | 120 1368 | 4445 | 6 |
| 55 | 9982 | 4459 | 8835 | 3055 | 7067 | 103 0819 | 4256 | 7327 | 5 |
| 56 | 016 2890 | 7366 | 051 1740 | 5957 | 9966 | 3712 | 7144 | 138 0208 | 4 |
| 57 | 5799 | 034 0274 | 4645 | 8859 | 086 2864 | 6605 | 121 0031 | 3089 | 3 |
| 58 | 8707 | 3181 | 7550 | 069 1761 | 5762 | 9499 | 2919 | 5970 | 2 |
| 59 | 017 1616 | 6088 | 052 0455 | 4663 | 8660 | 104 2392 | 5806 | 8850 | 1 |
| 60 | 4524 | 8995 | 3360 | 7565 | 087 1557 | 5285 | 8693 | 139 1731 | 0 |
| ′ | 89° | 88° | 87° | 86° | 85° | 84° | 83° | 82° | ′ |

NAT. COSINE.

| ′ | 0° | 1° | 2° | 3° | 4° | 5° | 6° | 7° | ′ |
|---|---|---|---|---|---|---|---|---|---|
| 0 | 000 0000 | 017 4551 | 034 9208 | 052 4078 | 069 9268 | 087 4887 | 105 1042 | 122 7840 | 60 |
| 1 | 2909 | 7460 | 035 2120 | 6995 | 070 2191 | 7818 | 3983 | 123 0798 | 59 |
| 2 | 5818 | 018 0370 | 5033 | 9912 | 5115 | 088 0749 | 6925 | 3752 | 58 |
| 3 | 8727 | 3280 | 7945 | 053 2829 | 8038 | 3681 | 9866 | 6705 | 57 |
| 4 | 001 1636 | 6190 | 036 0858 | 5746 | 071 0961 | 6612 | 106 2808 | 9658 | 56 |
| 5 | 4544 | 9100 | 3771 | 8663 | 3885 | 9544 | 5750 | 124 2612 | 55 |
| 6 | 7453 | 019 2010 | 6683 | 054 1581 | 6809 | 089 2476 | 8692 | 5566 | 54 |
| 7 | 002 0362 | 4920 | 9596 | 4498 | 9733 | 5408 | 107 1634 | 8520 | 53 |
| 8 | 3271 | 7830 | 037 2509 | 7416 | 072 2657 | 8341 | 4576 | 125 1474 | 52 |
| 9 | 6180 | 020 0740 | 5422 | 055 0333 | 5581 | 090 1273 | 7519 | 4429 | 51 |
| 10 | 9089 | 3650 | 8335 | 3251 | 8505 | 4206 | 108 0462 | 7384 | 50 |
| 11 | 003 1998 | 6560 | 038 1248 | 6169 | 073 1430 | 7138 | 3405 | 126 0339 | 49 |
| 12 | 4907 | 9470 | 4161 | 9087 | 4354 | 091 0071 | 6348 | 3294 | 48 |
| 13 | 7816 | 021 2380 | 7074 | 056 2005 | 7279 | 3004 | 9291 | 6243 | 47 |
| 14 | 004 0725 | 5291 | 9988 | 4923 | 074 0203 | 5938 | 109 2234 | 9205 | 46 |
| 15 | 3634 | 8201 | 039 2901 | 7841 | 3128 | 8871 | 5178 | 127 2161 | 45 |
| 16 | 6542 | 022 1111 | 5814 | 057 0759 | 6053 | 092 1804 | 8122 | 5117 | 44 |
| 17 | 9451 | 4021 | 8728 | 3678 | 8979 | 4738 | 110 1066 | 8073 | 43 |
| 18 | 005 2360 | 6932 | 040 1641 | 6596 | 075 1904 | 7672 | 4010 | 128 1030 | 42 |
| 19 | 5269 | 9842 | 4555 | 9515 | 4829 | 093 0606 | 6955 | 3986 | 41 |
| 20 | 8178 | 023 2753 | 7469 | 058 2434 | 7755 | 3540 | 9899 | 6945 | 40 |
| 21 | 006 1087 | 5663 | 041 0383 | 5352 | 076 0680 | 6474 | 111 2844 | 9900 | 39 |
| 22 | 3996 | 8574 | 3296 | 8271 | 3606 | 9409 | 5789 | 129 2858 | 38 |
| 23 | 6905 | 024 1484 | 6210 | 059 1190 | 6532 | 094 2344 | 8734 | 5815 | 37 |
| 24 | 9814 | 4395 | 9124 | 4109 | 9458 | 5278 | 112 1680 | 8773 | 36 |
| 25 | 007 2723 | 7305 | 042 2038 | 7029 | 077 2384 | 8213 | 4625 | 130 1731 | 35 |
| 26 | 5632 | 025 0216 | 4952 | 9948 | 5311 | 095 1148 | 7571 | 4690 | 34 |
| 27 | 8541 | 3127 | 7866 | 060 2867 | 8237 | 4084 | 113 0517 | 7648 | 33 |
| 28 | 008 1450 | 6038 | 043 0781 | 5787 | 078 1164 | 7019 | 3463 | 131 0607 | 32 |
| 29 | 4360 | 8948 | 3695 | 8706 | 4090 | 9955 | 6410 | 3566 | 31 |
| 30 | 7269 | 026 1859 | 6609 | 061 1626 | 7017 | 096 2890 | 9356 | 6525 | 30 |
| 31 | 009 0178 | 4770 | 9524 | 4546 | 9944 | 5826 | 114 2303 | 9484 | 29 |
| 32 | 3087 | 7681 | 044 2438 | 7466 | 079 2871 | 8763 | 5250 | 132 2444 | 28 |
| 33 | 5996 | 027 0592 | 5353 | 062 0386 | 5798 | 097 1699 | 8197 | 5404 | 27 |
| 34 | 8905 | 3503 | 8268 | 3306 | 8726 | 4635 | 115 1144 | 8364 | 26 |
| 35 | 010 1814 | 6414 | 045 1183 | 6226 | 080 1653 | 7572 | 4092 | 133 1324 | 25 |
| 36 | 4724 | 9325 | 4097 | 9147 | 4581 | 098 0509 | 7039 | 4285 | 24 |
| 37 | 7633 | 028 2236 | 7012 | 063 2067 | 7509 | 3446 | 9987 | 7246 | 23 |
| 38 | 011 0542 | 5148 | 9927 | 4988 | 081 0437 | 6383 | 116 2936 | 134 0207 | 22 |
| 39 | 3451 | 8059 | 046 2842 | 7908 | 3365 | 9320 | 5884 | 3168 | 21 |
| 40 | 6361 | 029 0970 | 5757 | 064 0829 | 6293 | 099 2257 | 8832 | 6129 | 20 |
| 41 | 9270 | 3882 | 8673 | 3750 | 9221 | 5194 | 117 1781 | 9091 | 19 |
| 42 | 012 2179 | 6793 | 047 1588 | 6671 | 082 2150 | 8133 | 4730 | 135 2053 | 18 |
| 43 | 5088 | 9705 | 4503 | 9592 | 5078 | 100 1071 | 7679 | 5015 | 17 |
| 44 | 7998 | 030 2616 | 7419 | 065 2513 | 8007 | 4009 | 118 0628 | 7978 | 16 |
| 45 | 013 0907 | 5528 | 048 0334 | 5435 | 083 0936 | 6947 | 3578 | 136 0940 | 15 |
| 46 | 3817 | 8439 | 3250 | 8356 | 3865 | 9886 | 6528 | 3903 | 14 |
| 47 | 6726 | 031 1351 | 6166 | 066 1278 | 6794 | 101 2824 | 9478 | 6866 | 13 |
| 48 | 9635 | 4263 | 9082 | 4199 | 9723 | 5763 | 119 2428 | 9830 | 12 |
| 49 | 014 2545 | 7174 | 049 1997 | 7121 | 084 2653 | 8702 | 5378 | 137 2793 | 11 |
| 50 | 5454 | 032 0086 | 4913 | 067 0043 | 5583 | 102 1641 | 8329 | 5757 | 10 |
| 51 | 8364 | 2998 | 7829 | 2965 | 8512 | 4580 | 120 1279 | 8721 | 9 |
| 52 | 015 1273 | 5910 | 050 0746 | 5887 | 085 1442 | 7520 | 4230 | 138 1685 | 8 |
| 53 | 4183 | 8822 | 3662 | 8809 | 4372 | 103 0460 | 7182 | 4650 | 7 |
| 54 | 7093 | 033 1734 | 6578 | 068 1732 | 7302 | 3399 | 121 0133 | 7615 | 6 |
| 55 | 016 0002 | 4646 | 9495 | 4654 | 086 0233 | 6340 | 3085 | 139 0580 | 5 |
| 56 | 2912 | 7558 | 051 2411 | 7577 | 3163 | 9280 | 6036 | 3545 | 4 |
| 57 | 5821 | 034 0471 | 5328 | 069 0499 | 6094 | 104 2220 | 8988 | 6510 | 3 |
| 58 | 8731 | 3383 | 8244 | 3422 | 9025 | 5161 | 122 1941 | 9476 | 2 |
| 59 | 017 1641 | 6295 | 052 1161 | 6345 | 087 1956 | 8101 | 4893 | 140 2442 | 1 |
| 60 | 4551 | 9208 | 4078 | 9268 | 4887 | 105 1042 | 7846 | 5408 | 0 |
| ′ | 89° | 88° | 87° | 86° | 85° | 84° | 83° | 82° | ′ |

NAT. COTAN.

| ′ | 8° | 9° | 10° | 11° | 12° | 13° | 14° | 15° | ′ |
|---|---|---|---|---|---|---|---|---|---|
| 0 | 139 1731 | 156 4345 | 173 6482 | 190 8090 | 207 9117 | 224 9511 | 241 9219 | 258 8190 | 60 |
| 1 | 4612 | 7218 | 9346 | 191 0945 | 208 1962 | 225 2345 | 242 2041 | 259 1006 | 59 |
| 2 | 7492 | 157 0091 | 174 2211 | 3801 | 4807 | 5179 | 4863 | 3810 | 58 |
| 3 | 140 0372 | 2963 | 5075 | 6656 | 7652 | 8013 | 7685 | 6619 | 57 |
| 4 | 3252 | 5836 | 7939 | 9510 | 209 0497 | 226 0846 | 243 0507 | 9428 | 56 |
| 5 | 6132 | 8708 | 175 0803 | 192 2365 | 3341 | 3680 | 3329 | 260 2237 | 55 |
| 6 | 9012 | 158 1581 | 3667 | 5220 | 6186 | 6513 | 6150 | 5045 | 54 |
| 7 | 141 1892 | 4453 | 6531 | 8074 | 9030 | 9346 | 8971 | 7853 | 53 |
| 8 | 4772 | 7325 | 9395 | 193 0928 | 210 1874 | 227 2179 | 244 1792 | 261 0662 | 52 |
| 9 | 7651 | 159 0197 | 176 2258 | 3782 | 4718 | 5012 | 4613 | 3469 | 51 |
| 10 | 142 0531 | 3069 | 5121 | 6636 | 7561 | 7844 | 7433 | 6277 | 50 |
| 11 | 3410 | 5940 | 7984 | 9490 | 211 0405 | 228 0677 | 245 0254 | 9085 | 49 |
| 12 | 6289 | 8812 | 177 0847 | 194 2344 | 3248 | 3509 | 3074 | 262 1892 | 48 |
| 13 | 9168 | 160 1683 | 3710 | 5197 | 6091 | 6341 | 5894 | 4699 | 47 |
| 14 | 143 2047 | 4555 | 6573 | 8050 | 8934 | 9172 | 8713 | 7506 | 46 |
| 15 | 4926 | 7426 | 9435 | 195 0903 | 212 1777 | 229 2004 | 246 1533 | 263 0312 | 45 |
| 16 | 7805 | 161 0297 | 178 2298 | 3756 | 4619 | 4835 | 4352 | 3118 | 44 |
| 17 | 144 0684 | 3167 | 5160 | 6609 | 7462 | 7666 | 7171 | 5925 | 43 |
| 18 | 3562 | 6038 | 8022 | 9461 | 213 0304 | 230 0497 | 9990 | 8730 | 42 |
| 19 | 6440 | 8909 | 179 0884 | 196 2314 | 3146 | 3328 | 247 2809 | 264 1536 | 41 |
| 20 | 9319 | 162 1779 | 3746 | 5166 | 5988 | 6159 | 5627 | 4342 | 40 |
| 21 | 145 2197 | 4650 | 6607 | 8018 | 8829 | 8989 | 8445 | 7147 | 39 |
| 22 | 5075 | 7520 | 9469 | 197 0870 | 214 1671 | 231 1819 | 248 1263 | 9952 | 38 |
| 23 | 7953 | 163 0390 | 180 2330 | 3722 | 4512 | 4649 | 4081 | 265 2757 | 37 |
| 24 | 146 0830 | 3260 | 5191 | 6573 | 7353 | 7479 | 6899 | 5561 | 36 |
| 25 | 3708 | 6129 | 8052 | 9425 | 215 0194 | 232 0309 | 9716 | 8366 | 35 |
| 26 | 6585 | 8999 | 181 0913 | 198 2276 | 3035 | 3138 | 249 2533 | 266 1170 | 34 |
| 27 | 9463 | 164 1868 | 3774 | 5127 | 5876 | 5967 | 5350 | 3973 | 33 |
| 28 | 147 2340 | 4738 | 6635 | 7978 | 8716 | 8796 | 8167 | 6777 | 32 |
| 29 | 5217 | 7607 | 9495 | 199 0829 | 216 1556 | 233 1625 | 250 0984 | 9581 | 31 |
| 30 | 8094 | 165 0476 | 182 2355 | 3679 | 4396 | 4454 | 3800 | 267 2384 | 30 |
| 31 | 148 0971 | 3345 | 5215 | 6530 | 7236 | 7282 | 6616 | 5187 | 29 |
| 32 | 3848 | 6214 | 8075 | 9380 | 217 0076 | 234 0110 | 9432 | 7989 | 28 |
| 33 | 6724 | 9082 | 183 0935 | 200 2230 | 2915 | 2938 | 251 2248 | 268 0792 | 27 |
| 34 | 9601 | 166 1951 | 3795 | 5080 | 5754 | 5766 | 5063 | 3594 | 26 |
| 35 | 149 2477 | 4819 | 6654 | 7930 | 8593 | 8594 | 7879 | 6396 | 25 |
| 36 | 5353 | 7687 | 9514 | 201 0779 | 218 1432 | 235 1421 | 252 0694 | 9198 | 24 |
| 37 | 8230 | 167 0556 | 184 2373 | 3629 | 4271 | 4248 | 3508 | 269 2000 | 23 |
| 38 | 150 1106 | 3423 | 5232 | 6478 | 7110 | 7075 | 6323 | 4801 | 22 |
| 39 | 3981 | 6291 | 8091 | 9327 | 9948 | 9902 | 9137 | 7602 | 21 |
| 40 | 6857 | 9159 | 185 0949 | 202 2176 | 219 2786 | 236 2729 | 253 1952 | 270 0403 | 20 |
| 41 | 9733 | 168 2026 | 3808 | 5024 | 5624 | 5555 | 4766 | 3204 | 19 |
| 42 | 151 2608 | 4894 | 6666 | 7873 | 8462 | 8381 | 7579 | 6004 | 18 |
| 43 | 5484 | 7761 | 9524 | 203 0721 | 220 1300 | 237 1207 | 254 0393 | 8805 | 17 |
| 44 | 8359 | 169 0628 | 186 2382 | 3569 | 4137 | 4033 | 3206 | 271 1605 | 16 |
| 45 | 152 1234 | 3495 | 5240 | 6418 | 6974 | 6859 | 6019 | 4404 | 15 |
| 46 | 4109 | 6362 | 8098 | 9265 | 9811 | 9684 | 8832 | 7204 | 14 |
| 47 | 6984 | 9228 | 187 0956 | 204 2113 | 221 2648 | 238 2510 | 255 1645 | 272 0003 | 13 |
| 48 | 9858 | 170 2095 | 3813 | 4961 | 5485 | 5335 | 4458 | 2802 | 12 |
| 49 | 153 2733 | 4961 | 6670 | 7808 | 8321 | 8159 | 7270 | 5601 | 11 |
| 50 | 5607 | 7828 | 9528 | 205 0655 | 222 1158 | 239 0984 | 256 0082 | 8400 | 10 |
| 51 | 8482 | 171 0694 | 188 2385 | 3502 | 3994 | 3808 | 2894 | 273 1198 | 9 |
| 52 | 154 1356 | 3560 | 5241 | 6349 | 6830 | 6633 | 5705 | 3997 | 8 |
| 53 | 4230 | 6425 | 8098 | 9195 | 9666 | 9457 | 8517 | 6794 | 7 |
| 54 | 7104 | 9291 | 189 0954 | 206 2042 | 223 2501 | 240 2280 | 257 1328 | 9592 | 6 |
| 55 | 9978 | 172 2156 | 3811 | 4888 | 5337 | 5104 | 4139 | 274 2390 | 5 |
| 56 | 155 2851 | 5022 | 6667 | 7734 | 8172 | 7927 | 6950 | 5187 | 4 |
| 57 | 5725 | 7887 | 9523 | 207 0580 | 224 1007 | 241 0751 | 9760 | 7984 | 3 |
| 58 | 8598 | 173 0752 | 190 2379 | 3426 | 3842 | 3574 | 258 2570 | 275 0781 | 2 |
| 59 | 156 1472 | 3617 | 5234 | 6272 | 6676 | 6396 | 5381 | 3577 | 1 |
| 60 | 4345 | 6482 | 8090 | 9117 | 9511 | 9219 | 8190 | 6374 | 0 |
| ′ | 81° | 80° | 79° | 78° | 77° | 76° | 75° | 74° | ′ |

NAT. COSINE.

| ′ | 8° | 9° | 10° | 11° | 12° | 13° | 14° | 15° | ′ |
|---|---|---|---|---|---|---|---|---|---|
| 0 | 140 5408 | 158 3844 | 176 3270 | 194 3803 | 212 5566 | 230 8682 | 249 3280 | 267 9492 | 60 |
| 1 | 8375 | 6826 | 6269 | 6822 | 8606 | 231 1746 | 6370 | 268 2610 | 59 |
| 2 | 141 1342 | 9809 | 9269 | 9841 | 213 1647 | 4811 | 9460 | 5728 | 58 |
| 3 | 4308 | 159 2791 | 177 2269 | 195 2861 | 4688 | 7876 | 250 2551 | 8847 | 57 |
| 4 | 7276 | 5774 | 5270 | 5881 | 7730 | 232 0941 | 5642 | 269 1967 | 56 |
| 5 | 142 0243 | 8757 | 8270 | 8901 | 214 0772 | 4007 | 8734 | 5087 | 55 |
| 6 | 3211 | 160 1740 | 178 1271 | 196 1922 | 3814 | 7073 | 251 1826 | 8207 | 54 |
| 7 | 6179 | 4724 | 4273 | 4943 | 6857 | 233 0140 | 4919 | 270 1328 | 53 |
| 8 | 9147 | 7708 | 7274 | 7964 | 9900 | 3207 | 8012 | 4449 | 52 |
| 9 | 143 2115 | 161 0692 | 179 0276 | 197 0986 | 215 2944 | 6274 | 252 1106 | 7571 | 51 |
| 10 | 5084 | 3677 | 3279 | 4008 | 5988 | 9342 | 4200 | 271 0694 | 50 |
| 11 | 8053 | 6662 | 6281 | 7031 | 9032 | 234 2410 | 7294 | 3817 | 49 |
| 12 | 144 1022 | 9647 | 9284 | 198 0053 | 216 2077 | 5479 | 253 0389 | 6940 | 48 |
| 13 | 3991 | 162 2632 | 180 2287 | 3076 | 5122 | 8548 | 3484 | 272 0064 | 47 |
| 14 | 6961 | 5618 | 5291 | 6100 | 8167 | 235 1617 | 6580 | 3188 | 46 |
| 15 | 9931 | 8603 | 8295 | 9124 | 217 1213 | 4687 | 9676 | 6313 | 45 |
| 16 | 145 2901 | 163 1590 | 181 1299 | 199 2148 | 4259 | 7758 | 254 2773 | 9438 | 44 |
| 17 | 5872 | 4576 | 4303 | 5172 | 7306 | 236 0829 | 5870 | 273 2564 | 43 |
| 18 | 8842 | 7563 | 7308 | 8197 | 218 0353 | 3900 | 8968 | 5690 | 42 |
| 19 | 146 1813 | 164 0550 | 182 0313 | 200 1222 | 3400 | 6971 | 255 2066 | 8817 | 41 |
| 20 | 4784 | 3537 | 3319 | 4248 | 6448 | 237 0044 | 5165 | 274 1945 | 40 |
| 21 | 7756 | 6525 | 6324 | 7274 | 9496 | 3116 | 8264 | 5072 | 39 |
| 22 | 147 0727 | 9513 | 9330 | 201 0300 | 219 2544 | 6189 | 256 1363 | 8201 | 38 |
| 23 | 3699 | 165 2501 | 183 2337 | 3327 | 5593 | 9262 | 4463 | 275 1330 | 37 |
| 24 | 6672 | 5489 | 5343 | 6354 | 8643 | 238 2336 | 7564 | 4459 | 36 |
| 25 | 9644 | 8478 | 8350 | 9381 | 220 1692 | 5410 | 257 0664 | 7589 | 35 |
| 26 | 148 2617 | 166 1467 | 184 1358 | 202 2409 | 4742 | 8465 | 3766 | 276 0719 | 34 |
| 27 | 5590 | 4456 | 4365 | 5437 | 7793 | 239 1560 | 6868 | 3850 | 33 |
| 28 | 8563 | 7446 | 7373 | 8465 | 221 0844 | 4635 | 9970 | 6981 | 32 |
| 29 | 149 1536 | 167 0436 | 185 0382 | 203 1494 | 3895 | 7711 | 258 3073 | 277 0113 | 31 |
| 30 | 4510 | 3426 | 3390 | 4523 | 6947 | 240 0788 | 6176 | 3245 | 30 |
| 31 | 7484 | 6417 | 6399 | 7552 | 9999 | 3864 | 9280 | 6378 | 29 |
| 32 | 150 0458 | 9407 | 9409 | 204 0582 | 222 3051 | 6942 | 259 2384 | 9512 | 28 |
| 33 | 3433 | 168 2398 | 186 2418 | 3612 | 6104 | 241 0019 | 5488 | 278 2646 | 27 |
| 34 | 6408 | 5390 | 5428 | 6643 | 9157 | 3097 | 8593 | 5780 | 26 |
| 35 | 9383 | 8381 | 8439 | 9674 | 223 2211 | 6176 | 260 1699 | 8915 | 25 |
| 36 | 151 2358 | 169 1373 | 187 1449 | 205 2705 | 5265 | 9255 | 4805 | 279 2050 | 24 |
| 37 | 5333 | 4366 | 4460 | 5737 | 8319 | 242 2334 | 7911 | 5186 | 23 |
| 38 | 8309 | 7358 | 7471 | 8769 | 224 1374 | 5414 | 261 1018 | 8322 | 22 |
| 39 | 152 1285 | 170 0351 | 188 0483 | 206 1801 | 4429 | 8494 | 4126 | 280 1459 | 21 |
| 40 | 4262 | 3344 | 3495 | 4834 | 7485 | 243 1575 | 7234 | 4597 | 20 |
| 41 | 7238 | 6338 | 6507 | 7867 | 225 0541 | 4656 | 262 0342 | 7735 | 19 |
| 42 | 153 0215 | 9331 | 9520 | 207 0900 | 3597 | 7737 | 3451 | 281 0873 | 18 |
| 43 | 3192 | 171 2325 | 189 2533 | 3934 | 6654 | 244 0819 | 6560 | 4012 | 17 |
| 44 | 6170 | 5320 | 5546 | 6968 | 9711 | 3902 | 9670 | 7152 | 16 |
| 45 | 9147 | 8314 | 8559 | 208 0003 | 226 2769 | 6984 | 263 2780 | 282 0292 | 15 |
| 46 | 154 2125 | 172 1309 | 190 1573 | 3038 | 5827 | 245 0068 | 5891 | 3432 | 14 |
| 47 | 5103 | 4304 | 4587 | 6073 | 8885 | 3151 | 9002 | 6573 | 13 |
| 48 | 8082 | 7300 | 7602 | 9109 | 227 1944 | 6236 | 264 2114 | 9715 | 12 |
| 49 | 155 1061 | 173 0296 | 191 0617 | 209 2145 | 5003 | 9320 | 5226 | 283 2857 | 11 |
| 50 | 4040 | 3292 | 3632 | 5181 | 8063 | 246 2405 | 8339 | 5999 | 10 |
| 51 | 7019 | 6288 | 6648 | 8218 | 228 1123 | 5491 | 265 1452 | 9143 | 9 |
| 52 | 9998 | 9285 | 9664 | 210 1255 | 4184 | 8577 | 4566 | 284 2286 | 8 |
| 53 | 156 2978 | 174 2282 | 192 2680 | 4293 | 7244 | 247 1663 | 7680 | 5430 | 7 |
| 54 | 5958 | 5279 | 5696 | 7331 | 229 0306 | 4750 | 266 0794 | 8575 | 6 |
| 55 | 8939 | 8277 | 8713 | 211 0369 | 3367 | 7837 | 3909 | 285 1720 | 5 |
| 56 | 157 1919 | 175 1275 | 193 1731 | 3407 | 6429 | 248 0925 | 7025 | 4866 | 4 |
| 57 | 4900 | 4273 | 4748 | 6446 | 9492 | 4013 | 267 0141 | 8012 | 3 |
| 58 | 7881 | 7272 | 7766 | 9486 | 230 2555 | 7102 | 3257 | 286 1159 | 2 |
| 59 | 158 0863 | 176 0271 | 194 0784 | 212 2525 | 5618 | 249 0191 | 6374 | 4306 | 1 |
| 60 | 3844 | 3270 | 3803 | 5566 | 8682 | 3280 | 9492 | 7454 | 0 |
| ′ | 81° | 80° | 79° | 78° | 77° | 76° | 75° | 74° | |

NAT. COTAN.

| ′ | 16° | 17° | 18° | 19° | 20° | 21° | 22° | 23° | ′ |
|---|---|---|---|---|---|---|---|---|---|
| 0 | 275 6374 | 292 3717 | 309 0170 | 325 5682 | 342 0201 | 358 3679 | 374 6066 | 390 7311 | 60 |
| 1 | 9170 | 6499 | 2936 | 8432 | 2935 | 6395 | 8763 | 9989 | 59 |
| 2 | 276 1965 | 9280 | 5702 | 326 1182 | 5668 | 9110 | 375 1459 | 391 2666 | 58 |
| 3 | 4761 | 293 2061 | 8468 | 3932 | 8400 | 359 1825 | 4156 | 5343 | 57 |
| 4 | 7556 | 4842 | 310 1234 | 6681 | 343 1133 | 4540 | 6852 | 8019 | 56 |
| 5 | 277 0352 | 7623 | 3999 | 9430 | 3865 | 7254 | 9547 | 392 0695 | 55 |
| 6 | 3147 | 294 0403 | 6764 | 327 2179 | 6597 | 9968 | 376 2243 | 3371 | 54 |
| 7 | 5941 | 3183 | 9529 | 4928 | 9329 | 360 2682 | 4938 | 6047 | 53 |
| 8 | 8736 | 5963 | 311 2294 | 7676 | 344 2060 | 5395 | 7632 | 8722 | 52 |
| 9 | 278 1530 | 8743 | 5058 | 328 0424 | 4791 | 8108 | 377 0327 | 393 1397 | 51 |
| 10 | 4324 | 295 1522 | 7822 | 3172 | 7521 | 361 0821 | 3021 | 4071 | 50 |
| 11 | 7118 | 4302 | 312 0586 | 5919 | 345 0252 | 3534 | 5714 | 6745 | 49 |
| 12 | 9911 | 7081 | 3349 | 8666 | 2982 | 6246 | 8408 | 9419 | 48 |
| 13 | 279 2704 | 9859 | 6112 | 329 1413 | 5712 | 8958 | 378 1101 | 394 2093 | 47 |
| 14 | 5497 | 296 2638 | 8875 | 4160 | 8441 | 362 1669 | 3794 | 4766 | 46 |
| 15 | 8290 | 5416 | 313 1638 | 6906 | 346 1171 | 4380 | 6486 | 7439 | 45 |
| 16 | 280 1083 | 8194 | 4400 | 9653 | 3900 | 7091 | 9178 | 395 0111 | 44 |
| 17 | 3875 | 297 0971 | 7163 | 320 2398 | 6628 | 9802 | 379 1870 | 2783 | 43 |
| 18 | 6667 | 3749 | 9925 | 5144 | 9357 | 363 2512 | 4562 | 5455 | 42 |
| 19 | 9459 | 6526 | 314 2686 | 7889 | 347 2085 | 5222 | 7253 | 8127 | 41 |
| 20 | 281 2251 | 9303 | 5448 | 331 0634 | 4812 | 7932 | 9944 | 396 0798 | 40 |
| 21 | 5042 | 298 2079 | 8209 | 3379 | 7540 | 364 0641 | 380 2634 | 3468 | 39 |
| 22 | 7833 | 4856 | 315 0969 | 6123 | 348 0267 | 3351 | 5324 | 6139 | 38 |
| 23 | 282 0624 | 7632 | 3730 | 8867 | 2994 | 6059 | 8014 | 8809 | 37 |
| 24 | 3415 | 299 0408 | 6490 | 332 1611 | 5720 | 8768 | 381 0704 | 397 1479 | 36 |
| 25 | 6205 | 3184 | 9250 | 4355 | 8447 | 365 1476 | 3393 | 4148 | 35 |
| 26 | 8995 | 5959 | 316 2010 | 7098 | 349 1173 | 4184 | 6082 | 6818 | 34 |
| 27 | 283 1785 | 8734 | 4770 | 9841 | 3898 | 6891 | 8770 | 9486 | 33 |
| 28 | 4575 | 300 1509 | 7529 | 333 2584 | 6624 | 9599 | 382 1459 | 398 2155 | 32 |
| 29 | 7364 | 4284 | 317 0288 | 5326 | 9349 | 366 2306 | 4147 | 4823 | 31 |
| 30 | 284 0153 | 7058 | 3047 | 8069 | 350 2074 | 5012 | 6834 | 7491 | 30 |
| 31 | 2942 | 9832 | 5805 | 334 0810 | 4798 | 7719 | 9522 | 399 0158 | 29 |
| 32 | 5731 | 301 2606 | 8563 | 3552 | 7523 | 367 0425 | 383 2209 | 2825 | 28 |
| 33 | 8520 | 5380 | 318 1321 | 6293 | 351 0246 | 3130 | 4895 | 5492 | 27 |
| 34 | 285 1308 | 8153 | 4079 | 9034 | 2970 | 5836 | 7582 | 8158 | 26 |
| 35 | 4096 | 302 0926 | 6836 | 335 1775 | 5963 | 8541 | 384 0268 | 400 0825 | 25 |
| 36 | 6884 | 3699 | 9593 | 4516 | 8416 | 368 1246 | 2953 | 3490 | 24 |
| 37 | 9671 | 6471 | 319 2350 | 7256 | 352 1139 | 3950 | 5639 | 6156 | 23 |
| 38 | 286 2458 | 9244 | 5106 | 9996 | 3862 | 6654 | 8324 | 8821 | 22 |
| 39 | 5246 | 303 2016 | 7863 | 336 2735 | 6584 | 9358 | 385 1008 | 401 1486 | 21 |
| 40 | 8032 | 4788 | 320 0619 | 5475 | 9306 | 369 2061 | 3693 | 4150 | 20 |
| 41 | 287 0819 | 7559 | 3374 | 8214 | 353 2027 | 4765 | 6377 | 6814 | 19 |
| 42 | 3605 | 304 0331 | 6130 | 337 0953 | 4748 | 7468 | 9060 | 9478 | 18 |
| 43 | 6391 | 3102 | 8885 | 3691 | 7469 | 370 0170 | 386 1744 | 402 2141 | 17 |
| 44 | 9177 | 5872 | 321 1640 | 6429 | 354 0190 | 2872 | 4427 | 4804 | 16 |
| 45 | 288 1963 | 8643 | 4395 | 9167 | 2910 | 5574 | 7110 | 7467 | 15 |
| 46 | 4748 | 305 1413 | 7149 | 338 1905 | 5630 | 8276 | 9792 | 403 0129 | 14 |
| 47 | 7533 | 4183 | 9903 | 4642 | 8350 | 371 0977 | 387 2474 | 2791 | 13 |
| 48 | 289 0318 | 6953 | 322 2657 | 7379 | 355 1070 | 3678 | 5156 | 5453 | 12 |
| 49 | 3103 | 9723 | 5411 | 339 0116 | 3789 | 6379 | 7837 | 8114 | 11 |
| 50 | 5887 | 306 2492 | 8164 | 2852 | 6508 | 9079 | 388 0518 | 404 0775 | 10 |
| 51 | 8671 | 5261 | 323 0917 | 5589 | 9226 | 372 1780 | 3199 | 3436 | 9 |
| 52 | 390 1455 | 8030 | 3670 | 8325 | 356 1944 | 4479 | 5880 | 6096 | 8 |
| 53 | 4239 | 307 0798 | 6422 | 340 1060 | 4662 | 7179 | 8560 | 8756 | 7 |
| 54 | 7022 | 3566 | 9174 | 3796 | 7380 | 9878 | 389 1240 | 405 1416 | 6 |
| 55 | 9805 | 6334 | 324 1926 | 6531 | 357 0097 | 373 2577 | 3919 | 4075 | 5 |
| 56 | 291 2588 | 9102 | 4678 | 9265 | 2814 | 5275 | 6598 | 6734 | 4 |
| 57 | 5371 | 308 1869 | 7429 | 341 2000 | 5531 | 7973 | 9277 | 9393 | 3 |
| 58 | 8153 | 4636 | 325 0180 | 4734 | 8248 | 374 0671 | 390 1955 | 406 2051 | 2 |
| 59 | 292 0935 | 7403 | 2931 | 7468 | 358 0964 | 3369 | 4633 | 4709 | 1 |
| 60 | 3717 | 309 0170 | 5682 | 342 0201 | 3679 | 6066 | 7311 | 7366 | 0 |
| ′ | 73° | 72° | 71° | 70° | 69° | 68° | 67° | 66° | ′ |

## NAT. TAN.

| ′ | 16° | 17° | 18° | 19° | 20° | 21° | 22° | 23° | ′ |
|---|---|---|---|---|---|---|---|---|---|
| 0 | 286 7454 | 305 7307 | 324 9197 | 344 3276 | 363 9702 | 383 8640 | 404 0262 | 424 4748 | 60 |
| 1 | 287 0602 | 306 0488 | 325 2413 | 6530 | 364 2997 | 384 1978 | 3646 | 8182 | 59 |
| 2 | 3751 | 3670 | 5630 | 9785 | 6292 | 5317 | 7031 | 425 1616 | 58 |
| 3 | 6900 | 6852 | 8848 | 345 3040 | 9588 | 8656 | 405 0417 | 5051 | 57 |
| 4 | 288 0050 | 307 0034 | 326 2066 | 6296 | 365 2885 | 385 1996 | 3804 | 8487 | 56 |
| 5 | 3201 | 3218 | 5284 | 9553 | 6182 | 5337 | 7191 | 426 1924 | 55 |
| 6 | 6352 | 6402 | 8504 | 346 2810 | 9480 | 8679 | 406 0579 | 5361 | 54 |
| 7 | 9503 | 9586 | 327 1724 | 6068 | 366 2779 | 386 2021 | 3968 | 8800 | 53 |
| 8 | 289 2655 | 308 2771 | 4944 | 9327 | 6079 | 5364 | 7358 | 427 2239 | 52 |
| 9 | 5808 | 5957 | 8165 | 347 2586 | 9379 | 8708 | 407 0748 | 5680 | 51 |
| 10 | 8961 | 9143 | 328 1387 | 5846 | 367 2680 | 387 2053 | 4139 | 9121 | 50 |
| 11 | 290 2114 | 309 2330 | 4610 | 9107 | 5981 | 5398 | 7531 | 428 2563 | 49 |
| 12 | 5269 | 5517 | 7833 | 348 2368 | 9284 | 8744 | 408 0924 | 6005 | 48 |
| 13 | 8423 | 8705 | 329 1056 | 5630 | 368 2587 | 388 2091 | 4318 | 9449 | 47 |
| 14 | 291 1578 | 310 1893 | 4281 | 8893 | 5890 | 5439 | 7713 | 429 2894 | 46 |
| 15 | 4734 | 5083 | 7505 | 349 2156 | 9195 | 8787 | 409 1108 | 6339 | 45 |
| 16 | 7890 | 8272 | 330 0731 | 5420 | 369 2500 | 389 2136 | 4504 | 9785 | 44 |
| 17 | 292 1047 | 311 1462 | 3957 | 8685 | 5806 | 5486 | 7901 | 430 3232 | 43 |
| 18 | 4205 | 4653 | 7184 | 350 1950 | 9112 | 8837 | 410 1299 | 6680 | 42 |
| 19 | 7363 | 7845 | 331 0411 | 5216 | 370 2420 | 390 2189 | 4697 | 431 0129 | 41 |
| 20 | 293 0521 | 312 1036 | 3639 | 8483 | 5728 | 5541 | 8097 | 3579 | 40 |
| 21 | 3680 | 4229 | 6868 | 351 1750 | 9036 | 8894 | 411 1497 | 7030 | 39 |
| 22 | 6839 | 7422 | 332 0097 | 5018 | 371 2346 | 391 2247 | 4898 | 432 0481 | 38 |
| 23 | 9999 | 313 0616 | 3327 | 8287 | 5656 | 5602 | 8300 | 3933 | 37 |
| 24 | 294 3160 | 3810 | 6557 | 352 1556 | 8967 | 8957 | 412 1703 | 7386 | 36 |
| 25 | 6321 | 7005 | 9788 | 4826 | 372 2278 | 392 2313 | 5106 | 433 0840 | 35 |
| 26 | 9483 | 314 0200 | 333 3020 | 8096 | 5590 | 5670 | 8510 | 4295 | 34 |
| 27 | 295 2645 | 3396 | 6252 | 353 1368 | 8903 | 9027 | 413 1915 | 7751 | 33 |
| 28 | 5808 | 6593 | 9485 | 4640 | 373 2217 | 393 2386 | 5321 | 434 1208 | 32 |
| 29 | 8971 | 9790 | 334 2719 | 7912 | 5532 | 5745 | 8728 | 4665 | 31 |
| 30 | 296 2135 | 315 2988 | 5953 | 354 1186 | 8847 | 9105 | 414 2136 | 8124 | 30 |
| 31 | 5209 | 6186 | 9188 | 4460 | 374 2163 | 394 2465 | 5544 | 435 1583 | 29 |
| 32 | 8464 | 9385 | 335 2424 | 7734 | 5479 | 5827 | 8953 | 5043 | 28 |
| 33 | 297 1630 | 316 2585 | 5660 | 355 1010 | 8797 | 9189 | 415 2363 | 8504 | 27 |
| 34 | 4796 | 5785 | 8896 | 4286 | 375 2115 | 395 2552 | 5774 | 436 1966 | 26 |
| 35 | 7962 | 8986 | 336 2134 | 7562 | 5433 | 5916 | 9186 | 5429 | 25 |
| 36 | 298 1129 | 317 2187 | 5372 | 356 0840 | 8753 | 9280 | 416 2598 | 8893 | 24 |
| 37 | 4297 | 5389 | 8610 | 4118 | 376 2073 | 396 2645 | 6012 | 437 2357 | 23 |
| 38 | 7465 | 8591 | 337 1850 | 7397 | 5394 | 6011 | 9426 | 5823 | 22 |
| 39 | 299 0634 | 318 1794 | 5090 | 357 0676 | 8716 | 9378 | 417 2841 | 9289 | 21 |
| 40 | 3803 | 4998 | 8330 | 3956 | 377 2038 | 397 2746 | 6257 | 438 2756 | 20 |
| 41 | 6973 | 8202 | 338 1571 | 7237 | 5361 | 6114 | 9673 | 6224 | 19 |
| 42 | 300 0144 | 319 1407 | 4813 | 358 0518 | 8685 | 9483 | 418 3091 | 9693 | 18 |
| 43 | 3315 | 4613 | 8056 | 3801 | 378 2010 | 398 2853 | 6509 | 439 3163 | 17 |
| 44 | 6486 | 7819 | 339 1299 | 7083 | 5335 | 6224 | 9928 | 6634 | 16 |
| 45 | 9658 | 320 1025 | 4543 | 359 0367 | 8661 | 9595 | 419 3348 | 440 0105 | 15 |
| 46 | 301 2831 | 4232 | 7787 | 3651 | 379 1988 | 399 2968 | 6769 | 3578 | 14 |
| 47 | 6004 | 7440 | 340 1032 | 6936 | 5315 | 6341 | 420 0190 | 7051 | 13 |
| 48 | 9178 | 321 0649 | 4278 | 360 0222 | 8644 | 9715 | 3613 | 441 0526 | 12 |
| 49 | 302 2352 | 3858 | 7524 | 3508 | 380 1973 | 400 3089 | 7036 | 4001 | 11 |
| 50 | 5527 | 7067 | 341 0771 | 6795 | 5302 | 6465 | 421 0460 | 7477 | 10 |
| 51 | 8703 | 322 0278 | 4019 | 361 0082 | 8633 | 9841 | 3885 | 442 0954 | 9 |
| 52 | 303 1879 | 3489 | 7267 | 3371 | 381 1964 | 401 3218 | 7311 | 4432 | 8 |
| 53 | 5055 | 6700 | 342 0516 | 6660 | 5296 | 6596 | 422 0738 | 7910 | 7 |
| 54 | 8232 | 9912 | 3765 | 9949 | 8629 | 9974 | 4165 | 443 1390 | 6 |
| 55 | 304 1410 | 323 3125 | 7015 | 362 3240 | 382 1962 | 402 3354 | 7594 | 4871 | 5 |
| 56 | 4588 | 6338 | 343 0266 | 6531 | 5296 | 6734 | 423 1023 | 8352 | 4 |
| 57 | 7767 | 9552 | 3518 | 9823 | 8631 | 403 0115 | 4453 | 444 1834 | 3 |
| 58 | 305 0946 | 324 2766 | 6770 | 363 3115 | 383 1967 | 3496 | 7884 | 5318 | 2 |
| 59 | 4126 | 5981 | 344 0023 | 6408 | 5303 | 6879 | 424 1316 | 8802 | 1 |
| 60 | 7307 | 9197 | 3276 | 9702 | 8640 | 404 0262 | 4748 | 445 2287 | 0 |
| ′ | 73° | 72° | 71° | 70° | 69° | 68° | 67° | 66° | ′ |

NAT. COTAN.

| ′ | 24° | 25° | 26° | 27° | 28° | 29° | 30° | 31° | ′ |
|---|---|---|---|---|---|---|---|---|---|
| 0 | 406 7366 | 422 6183 | 438 3711 | 453 9905 | 469 4716 | 484 8096 | 500 0000 | 515 0381 | 60 |
| 1 | 407 0024 | 8819 | 6326 | 454 2497 | 7284 | 485 0640 | 2519 | 2874 | 59 |
| 2 | 2681 | 423 1455 | 8940 | 5088 | 9852 | 3184 | 5037 | 5367 | 58 |
| 3 | 5337 | 4090 | 439 1553 | 7679 | 470 2419 | 5727 | 7556 | 7859 | 57 |
| 4 | 7993 | 6725 | 4166 | 455 0269 | 4986 | 8270 | 501 0073 | 516 0351 | 56 |
| 5 | 408 0649 | 9360 | 6779 | 2859 | 7553 | 486 0812 | 2591 | 2842 | 55 |
| 6 | 3305 | 424 1994 | 9392 | 5449 | 471 0119 | 3354 | 5107 | 5333 | 54 |
| 7 | 5960 | 4628 | 440 2004 | 8038 | 2685 | 5895 | 7624 | 7824 | 53 |
| 8 | 8615 | 7262 | 4615 | 456 0627 | 5250 | 8436 | 502 0140 | 517 0314 | 52 |
| 9 | 409 1269 | 9895 | 7227 | 3216 | 7815 | 487 0977 | 2655 | 2804 | 51 |
| 10 | 3923 | 425 2528 | 9838 | 5804 | 472 0380 | 3517 | 5170 | 5293 | 50 |
| 11 | 6577 | 5161 | 441 2448 | 8392 | 2944 | 6057 | 7685 | 7782 | 49 |
| 12 | 9230 | 7793 | 5059 | 457 0979 | 5508 | 8597 | 503 0199 | 518 0270 | 48 |
| 13 | 410 1883 | 426 0425 | 7668 | 3566 | 8071 | 488 1136 | 2713 | 2758 | 47 |
| 14 | 4536 | 3056 | 442 0278 | 6153 | 473 0634 | 3674 | 5227 | 5246 | 46 |
| 15 | 7189 | 5687 | 2887 | 8739 | 3197 | 6212 | 7740 | 7733 | 45 |
| 16 | 9841 | 8318 | 5496 | 458 1325 | 5759 | 8750 | 504 0252 | 519 0219 | 44 |
| 17 | 411 2492 | 427 0949 | 8104 | 3910 | 8321 | 489 1288 | 2765 | 2705 | 43 |
| 18 | 5144 | 3579 | 443 0712 | 6496 | 474 0882 | 3825 | 5276 | 5191 | 42 |
| 19 | 7795 | 6208 | 3319 | 9080 | 3443 | 6361 | 7788 | 7676 | 41 |
| 20 | 412 0445 | 8838 | 5927 | 459 1665 | 6004 | 8897 | 505 0298 | 520 0161 | 40 |
| 21 | 3096 | 428 1467 | 8534 | 4248 | 8564 | 490 1433 | 2809 | 2646 | 39 |
| 22 | 5745 | 4095 | 444 1140 | 6832 | 475 1124 | 3968 | 5319 | 5130 | 38 |
| 23 | 8395 | 6723 | 3746 | 9415 | 3683 | 6503 | 7828 | 7613 | 37 |
| 24 | 413 1044 | 9351 | 6352 | 460 1998 | 6242 | 9038 | 506 0338 | 521 0096 | 36 |
| 25 | 3693 | 429 1979 | 8957 | 4580 | 8801 | 491 1572 | 2846 | 2579 | 35 |
| 26 | 6342 | 4606 | 445 1562 | 7162 | 476 1359 | 4105 | 5355 | 5061 | 34 |
| 27 | 8990 | 7233 | 4167 | 9744 | 3917 | 6638 | 7863 | 7543 | 33 |
| 28 | 414 1638 | 9859 | 6771 | 461 2325 | 6474 | 9171 | 507 0370 | 522 0024 | 32 |
| 29 | 4285 | 430 2485 | 9375 | 4906 | 9031 | 492 1704 | 2877 | 2505 | 31 |
| 30 | 6932 | 5111 | 446 1978 | 7486 | 477 1588 | 4236 | 5384 | 4986 | 30 |
| 31 | 9579 | 7736 | 4581 | 462 0066 | 4144 | 6767 | 7890 | 7466 | 29 |
| 32 | 415 2226 | 431 0361 | 7184 | 2646 | 6700 | 9298 | 508 0396 | 9945 | 28 |
| 33 | 4872 | 2986 | 9786 | 5225 | 9255 | 493 1829 | 2901 | 523 2424 | 27 |
| 34 | 7517 | 5610 | 447 2388 | 7804 | 478 1810 | 4359 | 5406 | 4903 | 26 |
| 35 | 416 0163 | 8234 | 4990 | 463 0382 | 4364 | 6889 | 7910 | 7381 | 25 |
| 36 | 2808 | 432 0857 | 7591 | 2960 | 6919 | 9419 | 509 0414 | 9859 | 24 |
| 37 | 5453 | 3481 | 448 0192 | 5538 | 9472 | 494 1948 | 2918 | 524 2336 | 23 |
| 38 | 8097 | 6103 | 2792 | 8115 | 479 2026 | 4476 | 5421 | 4813 | 22 |
| 39 | 417 0741 | 8726 | 5392 | 464 0692 | 4579 | 7005 | 7924 | 7290 | 21 |
| 40 | 3385 | 433 1348 | 7992 | 3269 | 7131 | 9532 | 510 0426 | 9766 | 20 |
| 41 | 6028 | 3970 | 449 0591 | 5845 | 9683 | 495 2060 | 2928 | 525 2241 | 19 |
| 42 | 8671 | 6591 | 3190 | 8420 | 480 2235 | 4587 | 5429 | 4717 | 18 |
| 43 | 418 1313 | 9212 | 5789 | 465 0996 | 4786 | 7113 | 7930 | 7191 | 17 |
| 44 | 3956 | 434 1832 | 8387 | 3571 | 7337 | 9639 | 511 0431 | 9665 | 16 |
| 45 | 6597 | 4453 | 450 0984 | 6145 | 9888 | 496 2165 | 2931 | 526 2139 | 15 |
| 46 | 9239 | 7072 | 3582 | 8719 | 481 2438 | 4690 | 5431 | 4613 | 14 |
| 47 | 419 1880 | 9692 | 6179 | 466 1293 | 4987 | 7215 | 7930 | 7085 | 13 |
| 48 | 4521 | 435 2311 | 8775 | 3866 | 7537 | 9740 | 512 0429 | 9558 | 12 |
| 49 | 7161 | 4930 | 451 1372 | 6439 | 482 0086 | 497 2264 | 2927 | 527 2030 | 11 |
| 50 | 9801 | 7548 | 3967 | 9012 | 2634 | 4787 | 5425 | 4502 | 10 |
| 51 | 420 2441 | 436 0166 | 6563 | 467 1584 | 5182 | 7310 | 7923 | 6973 | 9 |
| 52 | 5080 | 2784 | 9158 | 4156 | 7730 | 9833 | 513 0420 | 9443 | 8 |
| 53 | 7719 | 5401 | 452 1753 | 6727 | 483 0277 | 498 2355 | 2916 | 528 1914 | 7 |
| 54 | 421 0358 | 8018 | 4347 | 9298 | 2824 | 4877 | 5413 | 4383 | 6 |
| 55 | 2996 | 437 0634 | 6941 | 468 1869 | 5370 | 7399 | 7908 | 6853 | 5 |
| 56 | 5634 | 3251 | 9535 | 4439 | 7916 | 9926 | 514 0404 | 9322 | 4 |
| 57 | 8272 | 5866 | 453 2128 | 7009 | 484 0462 | 499 2441 | 2899 | 529 1790 | 3 |
| 58 | 422 0909 | 8482 | 4721 | 9578 | 3007 | 4961 | 5393 | 4258 | 2 |
| 59 | 3546 | 438 1097 | 7313 | 469 2147 | 5552 | 7481 | 7887 | 6726 | 1 |
| 0 | 6183 | 3711 | 9905 | 4716 | 8096 | 500 0000 | 515 0381 | 9193 | 0 |
| ′ | 65° | 64° | 63° | 62° | 61° | 60° | 59° | 58° | ′ |

NAT. COSINE.

| ′ | 24° | 25° | 26° | 27° | 28° | 29° | 30° | 31° | ′ |
|---|---|---|---|---|---|---|---|---|---|
| 0 | 445 2287 | 466 3077 | 487 7326 | 509 5254 | 531 7094 | 554 3091 | 577 3503 | 600 8606 | 60 |
| 1 | 5773 | 6618 | 488 0927 | 8919 | 532 0826 | 6894 | 7382 | 601 2566 | 59 |
| 2 | 9260 | 467 0161 | 4530 | 510 2585 | 4559 | 555 0698 | 578 1262 | 6527 | 58 |
| 3 | 446 2747 | 3705 | 8133 | 6252 | 8293 | 4504 | 5144 | 602 0490 | 57 |
| 4 | 6236 | 7250 | 489 1737 | 9919 | 533 2029 | 8311 | 9027 | 4454 | 56 |
| 5 | 9726 | 468 0796 | 5343 | 511 3588 | 5765 | 556 2119 | 579 2912 | 8419 | 55 |
| 6 | 447 3216 | 4342 | 8949 | 7259 | 9503 | 5929 | 6797 | 603 2386 | 54 |
| 7 | 6708 | 7890 | 490 2557 | 512 0930 | 534 3242 | 9739 | 580 0684 | 6354 | 53 |
| 8 | 448 0200 | 469 1439 | 6166 | 4602 | 6981 | 557 3551 | 4573 | 604 0323 | 52 |
| 9 | 3693 | 4988 | 9775 | 8275 | 535 0723 | 7364 | 8462 | 4294 | 51 |
| 10 | 7187 | 8539 | 491 3386 | 513 1950 | 4465 | 558 1179 | 581 2353 | 8266 | 50 |
| 11 | 449 0682 | 470 2090 | 6997 | 5625 | 8208 | 4994 | 6245 | 605 2240 | 49 |
| 12 | 4178 | 5643 | 492 0610 | 9302 | 536 1953 | 8811 | 582 0139 | 6215 | 48 |
| 13 | 7675 | 9196 | 4224 | 514 2980 | 5699 | 559 2629 | 4034 | 606 0192 | 47 |
| 14 | 450 1173 | 471 2751 | 7838 | 6658 | 9446 | 6449 | 7930 | 4170 | 46 |
| 15 | 4672 | 6306 | 493 1454 | 515 0338 | 537 3194 | 560 0269 | 583 1828 | 8149 | 45 |
| 16 | 8171 | 9863 | 5071 | 4019 | 6943 | 4091 | 5726 | 607 2130 | 44 |
| 17 | 451 1672 | 472 3420 | 8689 | 7702 | 538 0694 | 7914 | 9627 | 6112 | 43 |
| 18 | 5173 | 6978 | 494 2308 | 516 1385 | 4445 | 561 1738 | 584 3528 | 608 0095 | 42 |
| 19 | 8676 | 473 0538 | 5928 | 5069 | 8198 | 5564 | 7431 | 4080 | 41 |
| 20 | 452 2179 | 4098 | 9549 | 8755 | 539 1952 | 9391 | 585 1335 | 8067 | 40 |
| 21 | 5683 | 7659 | 495 3171 | 517 2441 | 5707 | 562 3219 | 5241 | 609 2054 | 39 |
| 22 | 9188 | 474 1222 | 6794 | 6129 | 9464 | 7048 | 9148 | 6043 | 38 |
| 23 | 453 2694 | 4785 | 496 0418 | 9818 | 540 3221 | 563 0879 | 586 3056 | 610 0034 | 37 |
| 24 | 6201 | 8349 | 4043 | 518 3508 | 6980 | 4710 | 6965 | 4026 | 36 |
| 25 | 9709 | 475 1914 | 7669 | 7199 | 541 0740 | 8543 | 587 0876 | 8019 | 35 |
| 26 | 454 3218 | 5481 | 497 1297 | 519 0891 | 4501 | 564 2378 | 4788 | 611 2014 | 34 |
| 27 | 6728 | 9048 | 4925 | 4584 | 8263 | 6213 | 8702 | 6011 | 33 |
| 28 | 455 0238 | 476 2616 | 8554 | 8278 | 542 2027 | 565 0050 | 588 2616 | 612 0008 | 32 |
| 29 | 3750 | 6185 | 498 2185 | 520 1974 | 5791 | 3888 | 6533 | 4007 | 31 |
| 30 | 7263 | 9755 | 5816 | 5671 | 9557 | 7728 | 589 0450 | 8008 | 30 |
| 31 | 456 0776 | 477 3326 | 9449 | 9368 | 543 3324 | 566 1568 | 4369 | 613 2010 | 29 |
| 32 | 4290 | 6899 | 499 3082 | 521 3067 | 7092 | 5410 | 8289 | 6013 | 28 |
| 33 | 7806 | 478 0472 | 6717 | 6767 | 544 0862 | 9254 | 590 2211 | 614 0018 | 27 |
| 34 | 457 1322 | 4046 | 500 0352 | 522 0468 | 4632 | 567 3098 | 6134 | 4024 | 26 |
| 35 | 4839 | 7621 | 3989 | 4170 | 8404 | 6944 | 591 0058 | 8032 | 25 |
| 36 | 8357 | 479 1197 | 7627 | 7874 | 545 2177 | 568 0791 | 3984 | 615 2041 | 24 |
| 37 | 458 1877 | 4774 | 501 1266 | 523 1578 | 5951 | 4639 | 7910 | 6052 | 23 |
| 38 | 5397 | 8352 | 4906 | 5284 | 9727 | 8488 | 592 1839 | 616 0064 | 22 |
| 39 | 8918 | 480 1932 | 8547 | 8990 | 546 3503 | 569 2339 | 5768 | 4077 | 21 |
| 40 | 459 2439 | 5512 | 502 2189 | 524 2698 | 7281 | 6191 | 9699 | 8092 | 20 |
| 41 | 5962 | 9093 | 5832 | 6407 | 547 1060 | 570 0045 | 593 3632 | 617 2108 | 19 |
| 42 | 9486 | 481 2675 | 9476 | 525 0117 | 4840 | 3899 | 7565 | 6126 | 18 |
| 43 | 460 3011 | 6258 | 503 3121 | 3829 | 8621 | 7755 | 594 1501 | 618 0145 | 17 |
| 44 | 6537 | 9842 | 6768 | 7541 | 548 2404 | 571 1612 | 5437 | 4166 | 16 |
| 45 | 461 0063 | 482 3427 | 504 0415 | 526 1255 | 6188 | 5471 | 9375 | 8188 | 15 |
| 46 | 3591 | 7014 | 4063 | 4969 | 9973 | 9331 | 595 3314 | 619 2211 | 14 |
| 47 | 7119 | 483 0601 | 7713 | 8685 | 549 3759 | 572 3192 | 7255 | 6236 | 13 |
| 48 | 462 0649 | 4189 | 505 1363 | 527 2402 | 7547 | 7054 | 596 1196 | 620 0265 | 12 |
| 49 | 4179 | 7778 | 5015 | 6120 | 550 1335 | 573 0918 | 5140 | 4291 | 11 |
| 50 | 7710 | 484 1368 | 8668 | 9839 | 5125 | 4783 | 9084 | 8320 | 10 |
| 51 | 463 1243 | 4959 | 506 2322 | 528 3560 | 8916 | 8649 | 597 3030 | 621 2351 | 9 |
| 52 | 4776 | 8552 | 5977 | 7281 | 551 2708 | 574 2516 | 6978 | 6383 | 8 |
| 53 | 8310 | 485 2145 | 9633 | 529 1004 | 6502 | 6385 | 598 0926 | 622 0417 | 7 |
| 54 | 464 1845 | 5739 | 507 3290 | 4727 | 552 0297 | 575 0255 | 4877 | 4452 | 6 |
| 55 | 5382 | 9334 | 6948 | 8452 | 4093 | 4126 | 8828 | 8488 | 5 |
| 56 | 8919 | 486 2931 | 508 0607 | 530 2178 | 7890 | 7999 | 599 2781 | 623 2527 | 4 |
| 57 | 465 2457 | 6528 | 4267 | 5906 | 553 1688 | 576 1873 | 6735 | 6566 | 3 |
| 58 | 5996 | 487 0126 | 7929 | 9634 | 5488 | 5748 | 600 0691 | 624 0607 | 2 |
| 59 | 9536 | 3726 | 509 1591 | 531 3364 | 9288 | 9625 | 4648 | 4650 | 1 |
| 60 | 466 3077 | 7326 | 5254 | 7094 | 554 3091 | 577 3503 | 8606 | 8694 | 0 |
| ′ | 65° | 64° | 63° | 62° | 61° | 60° | 59° | 58° | ′ |

NAT. COTAN.

NAT. SINE.

| ′ | 32° | 33° | 34° | 35° | 36° | 37° | 38° | 39° | ′ |
|---|---|---|---|---|---|---|---|---|---|
| 0 | 529 9193 | 544 6390 | 559 1929 | 573 5764 | 587 7853 | 601 8150 | 615 6615 | 629 3204 | 60 |
| 1 | 530 1659 | 8830 | 4340 | 8147 | 588 0206 | 602 0473 | 8907 | 5464 | 59 |
| 2 | 4125 | 545 1269 | 6751 | 574 0529 | 2558 | 2795 | 616 1198 | 7724 | 58 |
| 3 | 6591 | 3707 | 9162 | 2911 | 4910 | 5117 | 3489 | 9983 | 57 |
| 4 | 9057 | 6145 | 560 1572 | 5292 | 7262 | 7439 | 5780 | 630 2242 | 56 |
| 5 | 531 1521 | 8583 | 3981 | 7672 | 9613 | 9760 | 8069 | 4500 | 55 |
| 6 | 3986 | 546 1020 | 6390 | 575 0053 | 589 1964 | 603 2080 | 617 0359 | 6758 | 54 |
| 7 | 6450 | 3456 | 8798 | 2432 | 4314 | 4400 | 2648 | 9015 | 53 |
| 8 | 8913 | 5892 | 561 1206 | 4811 | 6663 | 6719 | 4936 | 631 1272 | 52 |
| 9 | 532 1376 | 8328 | 3614 | 7190 | 9012 | 9038 | 7224 | 3528 | 51 |
| 10 | 3839 | 547 0763 | 6021 | 9568 | 590 1361 | 604 1356 | 9511 | 5784 | 50 |
| 11 | 6301 | 3198 | 8428 | 576 1946 | 3709 | 3674 | 618 1798 | 8039 | 49 |
| 12 | 8763 | 5632 | 562 0834 | 4323 | 6057 | 5991 | 4084 | 632 0293 | 48 |
| 13 | 533 1224 | 8066 | 3239 | 6700 | 8404 | 8308 | 6370 | 2547 | 47 |
| 14 | 3685 | 548 0499 | 5645 | 9076 | 591 0750 | 605 0624 | 8655 | 4800 | 46 |
| 15 | 6145 | 2932 | 8049 | 577 1452 | 3096 | 2940 | 619 0939 | 7053 | 45 |
| 16 | 8605 | 5365 | 563 0453 | 3827 | 5442 | 5255 | 3224 | 9306 | 44 |
| 17 | 534 1065 | 7797 | 2857 | 6202 | 7787 | 7570 | 5507 | 633 1557 | 43 |
| 18 | 3523 | 549 0228 | 5260 | 8576 | 592 0132 | 9884 | 7790 | 3809 | 42 |
| 19 | 5982 | 2659 | 7663 | 578 0950 | 2476 | 606 2198 | 620 0073 | 6059 | 41 |
| 20 | 8440 | 5090 | 564 0066 | 3323 | 4819 | 4511 | 2355 | 8310 | 40 |
| 21 | 535 0898 | 7520 | 2467 | 5696 | 7163 | 6824 | 4636 | 634 0559 | 39 |
| 22 | 3355 | 9950 | 4869 | 8069 | 9505 | 9136 | 6917 | 2808 | 38 |
| 23 | 5812 | 550 2379 | 7270 | 579 0440 | 593 1847 | 607 1447 | 9198 | 5057 | 37 |
| 24 | 8268 | 4807 | 9670 | 2812 | 4189 | 3758 | 621 1478 | 7305 | 36 |
| 25 | 536 0724 | 7236 | 565 2070 | 5183 | 6530 | 6069 | 3757 | 9553 | 35 |
| 26 | 3179 | 9663 | 4469 | 7553 | 8871 | 8379 | 6036 | 635 1800 | 34 |
| 27 | 5634 | 551 2091 | 6868 | 9923 | 594 1211 | 608 0689 | 8314 | 4046 | 33 |
| 28 | 8089 | 4518 | 9267 | 580 2292 | 3550 | 2998 | 622 0592 | 6292 | 32 |
| 29 | 537 0543 | 6944 | 566 1665 | 4661 | 5889 | 5306 | 2870 | 8537 | 31 |
| 30 | 2996 | 9370 | 4062 | 7030 | 8228 | 7614 | 5146 | 636 0782 | 30 |
| 31 | 5449 | 552 1795 | 6459 | 9397 | 595 0566 | 9922 | 7423 | 3026 | 29 |
| 32 | 7902 | 4220 | 8856 | 581 1765 | 2904 | 609 2229 | 9698 | 5270 | 28 |
| 33 | 538 0354 | 6645 | 567 1252 | 4132 | 5241 | 4535 | 623 1974 | 7513 | 27 |
| 34 | 2806 | 9069 | 3648 | 6498 | 7577 | 6841 | 4248 | 9756 | 26 |
| 35 | 5257 | 553 1492 | 6043 | 8864 | 9913 | 9147 | 6522 | 637 1998 | 25 |
| 36 | 7708 | 3915 | 8437 | 582 1230 | 596 2249 | 610 1452 | 8796 | 4240 | 24 |
| 37 | 539 0158 | 6338 | 568 0832 | 3595 | 4584 | 3756 | 624 1069 | 6481 | 23 |
| 38 | 2608 | 8760 | 3225 | 5959 | 6918 | 6060 | 3342 | 8721 | 22 |
| 39 | 5058 | 554 1182 | 5619 | 8323 | 9252 | 8363 | 5614 | 638 0961 | 21 |
| 40 | 7507 | 3603 | 8011 | 583 0687 | 597 1586 | 611 0666 | 7885 | 3201 | 20 |
| 41 | 9955 | 6024 | 569 0403 | 3050 | 3919 | 2969 | 625 0156 | 5440 | 19 |
| 42 | 540 2403 | 8444 | 2795 | 5412 | 6251 | 5270 | 2427 | 7678 | 18 |
| 43 | 4851 | 555 0864 | 5187 | 7774 | 8583 | 7572 | 4696 | 9916 | 17 |
| 44 | 7298 | 3283 | 7577 | 584 0136 | 598 0915 | 9873 | 6966 | 639 2153 | 16 |
| 45 | 9745 | 5702 | 9968 | 2497 | 3246 | 612 2173 | 9235 | 4390 | 15 |
| 46 | 541 2191 | 8121 | 570 2357 | 4857 | 5577 | 4473 | 626 1503 | 6626 | 14 |
| 47 | 4637 | 556 0539 | 4747 | 7217 | 7906 | 6772 | 3771 | 8862 | 13 |
| 48 | 7082 | 2956 | 7136 | 9577 | 599 0236 | 9071 | 6038 | 640 1097 | 12 |
| 49 | 9527 | 5373 | 9524 | 585 1936 | 2565 | 613 1369 | 8305 | 3332 | 11 |
| 50 | 542 1971 | 7790 | 571 1912 | 4294 | 4893 | 3666 | 627 0571 | 5566 | 10 |
| 51 | 4415 | 557 0206 | 4299 | 6652 | 7221 | 5964 | 2837 | 7799 | 9 |
| 52 | 6859 | 2621 | 6686 | 9010 | 9549 | 8260 | 5102 | 641 0032 | 8 |
| 53 | 9302 | 5036 | 9073 | 586 1367 | 600 1876 | 614 0556 | 7366 | 2264 | 7 |
| 54 | 543 1744 | 7451 | 572 1459 | 3724 | 4202 | 2852 | 9631 | 4496 | 6 |
| 55 | 4187 | 9865 | 3844 | 6080 | 6528 | 5147 | 628 1894 | 6728 | 5 |
| 56 | 6628 | 558 2279 | 6229 | 8435 | 8854 | 7442 | 4157 | 8958 | 4 |
| 57 | 9069 | 4692 | 8614 | 587 0790 | 601 1179 | 9736 | 6420 | 642 1189 | 3 |
| 58 | 544 1510 | 7105 | 573 0998 | 3145 | 3503 | 615 2029 | 8682 | 3418 | 2 |
| 59 | 3951 | 9517 | 3381 | 5499 | 5827 | 4322 | 629 0943 | 5647 | 1 |
| 60 | 6390 | 559 1929 | 5764 | 7853 | 8150 | 6615 | 3204 | 7876 | 0 |
| ′ | 57° | 56° | 55° | 54° | 53° | 52° | 51° | 50° | ′ |

NAT. COSINE.

| ′ | 32° | 33° | 34° | 35° | 36° | 37° | 38° | 39° | ′ |
|---|---|---|---|---|---|---|---|---|---|
| 0 | 624 8694 | 649 4076 | 674 5085 | 700 2075 | 726 5425 | 753 5541 | 781 2856 | 809 7840 | 60 |
| 1 | 625 2739 | 8212 | 9318 | 6411 | 9871 | 754 0102 | 7542 | 810 2658 | 59 |
| 2 | 6786 | 650 2350 | 675 3553 | 701 0749 | 727 4318 | 4666 | 782 2229 | 7478 | 58 |
| 3 | 626 0834 | 6490 | 7790 | 5089 | 8767 | 9232 | 6919 | 811 2300 | 57 |
| 4 | 4884 | 651 0631 | 676 2028 | 9430 | 728 3218 | 755 3799 | 783 1611 | 7124 | 56 |
| 5 | 8935 | 4774 | 6268 | 702 3773 | 7671 | 8369 | 6305 | 812 1951 | 55 |
| 6 | 627 2988 | 8918 | 677 0509 | 8118 | 729 2125 | 756 2941 | 784 1002 | 6780 | 54 |
| 7 | 7042 | 652 3064 | 4752 | 703 2464 | 6582 | 7514 | 5700 | 813 1611 | 53 |
| 8 | 628 1098 | 7211 | 8997 | 6813 | 730 1041 | 757 2090 | 785 0400 | 6444 | 52 |
| 9 | 5155 | 653 1360 | 678 3243 | 704 1163 | 5501 | 6668 | 5103 | 814 1280 | 51 |
| 10 | 9214 | 5511 | 7492 | 5515 | 9963 | 758 1248 | 9808 | 6118 | 50 |
| 11 | 629 3274 | 9663 | 679 1741 | 9869 | 731 4428 | 5829 | 786 4515 | 815 0958 | 49 |
| 12 | 7336 | 654 3817 | 5993 | 705 4224 | 8894 | 759 0413 | 9224 | 5801 | 48 |
| 13 | 630 1399 | 7972 | 680 0246 | 8581 | 732 3362 | 4999 | 787 3935 | 816 0646 | 47 |
| 14 | 5464 | 655 2129 | 4501 | 706 2940 | 7832 | 9587 | 8649 | 5493 | 46 |
| 15 | 9530 | 6287 | 8758 | 7301 | 733 2303 | 760 4177 | 788 3364 | 817 0343 | 45 |
| 16 | 631 3598 | 656 0447 | 681 3016 | 707 1664 | 6777 | 8769 | 8082 | 5195 | 44 |
| 17 | 7667 | 4609 | 7276 | 6028 | 734 1253 | 761 3363 | 789 2802 | 818 0049 | 43 |
| 18 | 632 1738 | 8772 | 682 1537 | 708 0395 | 5730 | 7959 | 7524 | 4905 | 42 |
| 19 | 5810 | 657 2937 | 5801 | 4763 | 735 0210 | 762 2557 | 790 2248 | 9764 | 41 |
| 20 | 9883 | 7103 | 683 0066 | 9133 | 4691 | 7157 | 6975 | 819 4625 | 40 |
| 21 | 633 3959 | 658 1271 | 4333 | 709 3504 | 9174 | 763 1759 | 791 1703 | 9488 | 39 |
| 22 | 8035 | 5441 | 8601 | 7878 | 736 3660 | 6363 | 6434 | 820 4354 | 38 |
| 23 | 634 2113 | 9612 | 684 2871 | 710 2253 | 8147 | 764 0969 | 792 1167 | 9222 | 37 |
| 24 | 6193 | 659 3785 | 7143 | 6630 | 737 2636 | 5577 | 5902 | 821 4093 | 36 |
| 25 | 635 0274 | 7960 | 685 1416 | 711 1009 | 7127 | 765 0188 | 793 0640 | 8965 | 35 |
| 26 | 4357 | 660 2136 | 5692 | 5390 | 738 1620 | 4800 | 5379 | 822 3840 | 34 |
| 27 | 8441 | 6313 | 9969 | 9772 | 6115 | 9414 | 794 0121 | 8718 | 33 |
| 28 | 636 2527 | 661 0492 | 686 4247 | 712 4157 | 739 0611 | 766 4031 | 4865 | 823 3597 | 32 |
| 29 | 6614 | 4673 | 8528 | 8543 | 5110 | 8649 | 9611 | 8479 | 31 |
| 30 | 637 0703 | 8856 | 687 2810 | 713 2931 | 9611 | 767 3270 | 795 4359 | 824 3364 | 30 |
| 31 | 4793 | 662 3040 | 7093 | 7320 | 740 4113 | 7893 | 9110 | 8251 | 29 |
| 32 | 8885 | 7225 | 688 1379 | 714 1712 | 8618 | 768 2517 | 796 3862 | 825 3140 | 28 |
| 33 | 638 2978 | 663 1413 | 5666 | 6106 | 741 3124 | 7144 | 8617 | 8031 | 27 |
| 34 | 7073 | 5601 | 9955 | 715 0501 | 7633 | 769 1773 | 797 3374 | 826 2925 | 26 |
| 35 | 639 1169 | 9792 | 689 4246 | 4898 | 742 2143 | 6404 | 8134 | 7821 | 25 |
| 36 | 5267 | 664 3984 | 8538 | 9297 | 6655 | 770 1037 | 798 2895 | 827 2719 | 24 |
| 37 | 9366 | 8178 | 690 2832 | 716 3698 | 743 1170 | 5672 | 7659 | 7620 | 23 |
| 38 | 640 3467 | 665 2373 | 7128 | 8100 | 5686 | 771 0309 | 799 2425 | 828 2523 | 22 |
| 39 | 7569 | 6570 | 691 1425 | 717 2505 | 744 0204 | 4948 | 7193 | 7429 | 21 |
| 40 | 641 1673 | 666 0769 | 5725 | 6911 | 4724 | 9589 | 800 1963 | 829 2337 | 20 |
| 41 | 5779 | 4969 | 692 0026 | 718 1319 | 9246 | 772 4233 | 6736 | 7247 | 19 |
| 42 | 9886 | 9171 | 4328 | 5729 | 745 3770 | 8878 | 801 1511 | 830 2160 | 18 |
| 43 | 642 3994 | 667 3374 | 8633 | 719 0141 | 8296 | 773 3526 | 6288 | 7075 | 17 |
| 44 | 8105 | 7580 | 693 2939 | 4554 | 746 2824 | 8176 | 802 1067 | 831 1992 | 16 |
| 45 | 643 2216 | 668 1786 | 7247 | 8970 | 7354 | 774 2827 | 5849 | 6912 | 15 |
| 46 | 6329 | 5995 | 694 1557 | 720 3387 | 747 1886 | 7481 | 803 0632 | 832 1834 | 14 |
| 47 | 644 0444 | 669 0205 | 5868 | 7806 | 6420 | 775 2137 | 5418 | 6759 | 13 |
| 48 | 4560 | 4417 | 695 0181 | 721 2227 | 748 0956 | 6795 | 804 0206 | 833 1686 | 12 |
| 49 | 8678 | 8630 | 4496 | 6650 | 5494 | 776 1455 | 4997 | 6615 | 11 |
| 50 | 645 2797 | 670 2845 | 8813 | 722 1075 | 749 0033 | 6118 | 9790 | 834 1547 | 10 |
| 51 | 6918 | 7061 | 696 3131 | 5502 | 4575 | 777 0782 | 805 4584 | 6481 | 9 |
| 52 | 646 1041 | 671 1280 | 7451 | 9930 | 9119 | 5448 | 9382 | 835 1418 | 8 |
| 53 | 5165 | 5500 | 697 1773 | 723 4361 | 750 3665 | 778 0117 | 806 4181 | 6357 | 7 |
| 54 | 9290 | 9721 | 6097 | 8793 | 8212 | 4788 | 8983 | 836 1298 | 6 |
| 55 | 647 3417 | 672 3944 | 698 0422 | 724 3227 | 751 2762 | 9460 | 807 3787 | 6242 | 5 |
| 56 | 7546 | 8169 | 4749 | 7663 | 7314 | 779 4135 | 8593 | 837 1188 | 4 |
| 57 | 648 1676 | 673 2396 | 9078 | 725 2101 | 752 1867 | 8812 | 808 3401 | 6136 | 3 |
| 58 | 5808 | 6624 | 699 3409 | 6540 | 6423 | 780 3492 | 8212 | 838 1087 | 2 |
| 59 | 9941 | 674 0854 | 7741 | 726 0982 | 753 0981 | 8173 | 809 3025 | 6041 | 1 |
| 60 | 649 4076 | 5085 | 700 2075 | 5425 | 5541 | 781 2856 | 7840 | 839 0996 | 0 |
| ′ | 57° | 56° | 55° | 54° | 53° | 52° | 51° | 50° | ′ |

| ′ | 40° | 41° | 42° | 43° | 44° | 45° | 46° | 47° | ′ |
|---|---|---|---|---|---|---|---|---|---|
| 0 | 642 7876 | 656 0590 | 669 1306 | 681 9984 | 694 6584 | 707 1068 | 719 3398 | 731 3537 | 60 |
| 1 | 643 0104 | 2785 | 3468 | 682 2111 | 8676 | 3124 | 5418 | 552. | 59 |
| 2 | 2332 | 4980 | 5628 | 4237 | 695 0767 | 5180 | 7438 | 7503 | 58 |
| 3 | 4559 | 7174 | 7789 | 6363 | 2858 | 7236 | 9457 | 9486 | 57 |
| 4 | 6785 | 9367 | 9948 | 8489 | 4949 | 9291 | 720 1476 | 732 1467 | 56 |
| 5 | 9011 | 657 1560 | 670 2108 | 683 0613 | 7039 | 708 1345 | 3494 | 3449 | 55 |
| 6 | 644 1236 | 3752 | 4266 | 2738 | 9128 | 3398 | 5511 | 5429 | 54 |
| 7 | 3461 | 5944 | 6424 | 4861 | 696 1217 | 5451 | 7528 | 7409 | 53 |
| 8 | 5685 | 8135 | 8582 | 6984 | 3305 | 7504 | 9544 | 9388 | 52 |
| 9 | 7909 | 658 0326 | 671 0739 | 9107 | 5392 | 9556 | 721 1559 | 733 1367 | 51 |
| 10 | 645 0132 | 2516 | 2895 | 684 1229 | 7479 | 709 1607 | 3574 | 3345 | 50 |
| 11 | 2355 | 4706 | 5051 | 3350 | 9565 | 3657 | 5589 | 5322 | 49 |
| 12 | 4577 | 6895 | 7206 | 5471 | 697 1651 | 5707 | 7602 | 7299 | 48 |
| 13 | 6798 | 9083 | 9361 | 7591 | 3736 | 7757 | 9615 | 9275 | 47 |
| 14 | 9019 | 659 1271 | 672 1515 | 9711 | 5821 | 9806 | 722 1628 | 734 1250 | 46 |
| 15 | 646 1240 | 3458 | 3668 | 685 1830 | 7905 | 710 1854 | 3640 | 3225 | 45 |
| 16 | 3460 | 5645 | 5821 | 3948 | 9988 | 3901 | 5651 | 5199 | 44 |
| 17 | 5679 | 7831 | 7973 | 6066 | 698 2071 | 5948 | 7661 | 7173 | 43 |
| 18 | 7898 | 660 0017 | 673 0125 | 8184 | 4153 | 7995 | 9671 | 9146 | 42 |
| 19 | 647 0116 | 2202 | 2276 | 686 0300 | 6234 | 711 0041 | 723 1681 | 735 1118 | 41 |
| 20 | 2334 | 4386 | 4427 | 2416 | 8315 | 2086 | 3690 | 3090 | 40 |
| 21 | 4551 | 6570 | 6577 | 4532 | 699 0396 | 4130 | 5698 | 5061 | 39 |
| 22 | 6767 | 8754 | 8727 | 6647 | 2476 | 6174 | 7705 | 7032 | 38 |
| 23 | 8984 | 661 0936 | 674 0876 | 8761 | 4555 | 8218 | 9712 | 9002 | 37 |
| 24 | 648 1199 | 3119 | 3024 | 687 0875 | 6633 | 712 0260 | 724 1719 | 736 0971 | 36 |
| 25 | 3414 | 5300 | 5172 | 2988 | 8711 | 2303 | 3724 | 2940 | 35 |
| 26 | 5628 | 7482 | 7319 | 5101 | 700 0789 | 4344 | 5729 | 4908 | 34 |
| 27 | 7842 | 9662 | 9466 | 7213 | 2866 | 6385 | 7734 | 6875 | 33 |
| 28 | 649 0056 | 662 1842 | 675 1612 | 9325 | 4942 | 8426 | 9738 | 8842 | 32 |
| 29 | 2268 | 4022 | 3757 | 688 1435 | 7018 | 713 0465 | 725 1741 | 737 0808 | 31 |
| 30 | 4480 | 6200 | 5902 | 3546 | 9093 | 2504 | 3744 | 2773 | 30 |
| 31 | 6692 | 8379 | 8046 | 5655 | 701 1167 | 4543 | 5746 | 4738 | 29 |
| 32 | 8903 | 663 0557 | 676 0190 | 7765 | 3241 | 6581 | 7747 | 6703 | 28 |
| 33 | 650 1114 | 2734 | 2333 | 9873 | 5314 | 8618 | 9748 | 8666 | 27 |
| 34 | 3324 | 4910 | 4476 | 689 1981 | 7387 | 714 0655 | 726 1748 | 738 0629 | 26 |
| 35 | 5533 | 7087 | 6618 | 4089 | 9459 | 2691 | 3748 | 2592 | 25 |
| 36 | 7742 | 9262 | 8760 | 6195 | 702 1531 | 4727 | 5747 | 4553 | 24 |
| 37 | 9951 | 664 1437 | 677 0901 | 8302 | 3601 | 6762 | 7745 | 6515 | 23 |
| 38 | 651 2158 | 3612 | 3041 | 690 0407 | 5672 | 8796 | 9743 | 8475 | 22 |
| 39 | 4366 | 5785 | 5181 | 2512 | 7741 | 715 0830 | 727 1740 | 739 0435 | 21 |
| 40 | 6572 | 7959 | 7320 | 4617 | 9811 | 2863 | 3736 | 2394 | 20 |
| 41 | 8778 | 665 0131 | 9459 | 6721 | 703 1879 | 4895 | 5732 | 4353 | 19 |
| 42 | 652 0984 | 2304 | 678 1597 | 8824 | 3947 | 6927 | 7728 | 6311 | 18 |
| 43 | 3189 | 4475 | 3734 | 691 0927 | 6014 | 8959 | 9722 | 8268 | 17 |
| 44 | 5394 | 6646 | 5871 | 3029 | 8081 | 716 0989 | 728 1716 | 740 0225 | 16 |
| 45 | 7598 | 8817 | 8007 | 5131 | 704 0147 | 3019 | 3710 | 2181 | 15 |
| 46 | 9801 | 666 0987 | 679 0143 | 7232 | 2213 | 5049 | 5703 | 4137 | 14 |
| 47 | 653 2004 | 3156 | 2278 | 9332 | 4278 | 7078 | 7695 | 6092 | 13 |
| 48 | 4206 | 5325 | 4413 | 692 1432 | 6342 | 9106 | 9686 | 8046 | 12 |
| 49 | 6408 | 7493 | 6547 | 3531 | 8406 | 717 1134 | 729 1677 | 741 0000 | 11 |
| 50 | 8609 | 9661 | 8681 | 5630 | 705 0469 | 3161 | 3668 | 1953 | 10 |
| 51 | 654 0810 | 667 1828 | 680 0813 | 7728 | 2532 | 5187 | 5657 | 3905 | 9 |
| 52 | 3010 | 3994 | 2946 | 9825 | 4594 | 7213 | 7646 | 5857 | 8 |
| 53 | 5209 | 6160 | 5078 | 693 1922 | 6655 | 9238 | 9635 | 7808 | 7 |
| 54 | 7408 | 8326 | 7209 | 4018 | 8716 | 718 1263 | 730 1623 | 9758 | 6 |
| 55 | 9607 | 668 0490 | 9339 | 6114 | 706 0776 | 3287 | 3610 | 742 1708 | 5 |
| 56 | 655 1804 | 2655 | 681 1469 | 8209 | 2835 | 5310 | 5597 | 3658 | 4 |
| 57 | 4002 | 4818 | 3599 | 694 0304 | 4894 | 7333 | 7583 | 5606 | 3 |
| 58 | 6198 | 6981 | 5728 | 2398 | 6953 | 9355 | 9568 | 7554 | 2 |
| 59 | 8395 | 9144 | 7856 | 4491 | 9011 | 719 1377 | 731 1553 | 9502 | 1 |
| 60 | 656 0590 | 669 1306 | 9984 | 6584 | 707 1068 | 3398 | 3537 | 743 1448 | 0 |
| | 49° | 48° | 47° | 46° | 45° | 44° | 43° | 42° | |

NAT. COSINE.

| ′ | 40° | 41° | 42° | 43° | 44° | 45° | 46° | 47° | ′ |
|---|---|---|---|---|---|---|---|---|---|
| 0 | 8390996 | 8692867 | 9004040 | 9325151 | 9656888 | 1·0000000 | 1·0355303 | 1·0723687 | 60 |
| 1 | 5955 | 7976 | 9309 | 9330591 | 9662511 | 05819 | 61333 | 29943 | 59 |
| 2 | 8400915 | 8703087 | 9014580 | 6034 | 8137 | 11642 | 67367 | 36203 | 58 |
| 3 | 5878 | 8200 | 9854 | 9341479 | 9673767 | 17469 | 73404 | 42467 | 57 |
| 4 | 8410844 | 8713316 | 9025131 | 6928 | 9399 | 23298 | 79445 | 48734 | 56 |
| 5 | 5812 | 8435 | 9030411 | 9352380 | 9685035 | 29131 | 85489 | 55006 | 55 |
| 6 | 8420782 | 8723556 | 5693 | 7834 | 9690674 | 34968 | 91538 | 61282 | 54 |
| 7 | 5755 | 8680 | 9040979 | 9363292 | 6316 | 40807 | 97589 | 67561 | 53 |
| 8 | 8430730 | 8733806 | 6267 | 8753 | 9701962 | 46651 | 1·0403645 | 73845 | 52 |
| 9 | 5708 | 8935 | 9051557 | 9374216 | 7610 | 52497 | 09704 | 80132 | 51 |
| 10 | 8440688 | 8744067 | 6851 | 9683 | 9713262 | 58348 | 15767 | 86423 | 50 |
| 11 | 5670 | 9201 | 9062147 | 9385153 | 8917 | 64201 | 21833 | 92718 | 49 |
| 12 | 8450655 | 8754338 | 7446 | 9390625 | 9724575 | 70058 | 27904 | 99018 | 48 |
| 13 | 5643 | 9478 | 9072748 | 6101 | 9730236 | 75918 | 33977 | 1·0805321 | 47 |
| 14 | 8460633 | 8764620 | 8053 | 9401579 | 5901 | 81782 | 40055 | 11628 | 46 |
| 15 | 5625 | 9765 | 9083360 | 7061 | 9741569 | 87649 | 46130 | 17939 | 45 |
| 16 | 8470620 | 8774912 | 8671 | 9412545 | 7240 | 93520 | 52221 | 24254 | 44 |
| 17 | 5617 | 8780062 | 9093984 | 8033 | 9752914 | 99394 | 58310 | 30573 | 43 |
| 18 | 8480617 | 5215 | 9300 | 9423523 | 8591 | 1·0105272 | 64402 | 36896 | 42 |
| 19 | 5619 | 8790370 | 9104919 | 9017 | 9764272 | 11153 | 70498 | 43223 | 41 |
| 20 | 8490624 | 5528 | 9940 | 9434513 | 9956 | 17038 | 76598 | 49554 | 40 |
| 21 | 5631 | 8800688 | 9115265 | 9440013 | 9775643 | 22925 | 82702 | 55889 | 39 |
| 22 | 8500640 | 5852 | 9120592 | 5516 | 9781333 | 28817 | 88809 | 62228 | 38 |
| 23 | 5653 | 8811017 | 5922 | 9451021 | 7027 | 34712 | 94920 | 68571 | 37 |
| 24 | 8510667 | 6186 | 9131255 | 6530 | 9792724 | 40610 | 1·0501034 | 74918 | 36 |
| 25 | 5684 | 8821357 | 6591 | 9462042 | 8424 | 46512 | 07153 | 81269 | 35 |
| 26 | 8520704 | 6531 | 9141929 | 7556 | 9804127 | 52418 | 13275 | 87624 | 34 |
| 27 | 5726 | 8831707 | 7270 | 9473074 | 9833 | 58326 | 19401 | 93984 | 33 |
| 28 | 8530750 | 6886 | 9152615 | 8595 | 9815543 | 64239 | 25531 | 1·0900347 | 32 |
| 29 | 5777 | 8842068 | 7962 | 9484119 | 9821256 | 70155 | 31664 | 06714 | 31 |
| 30 | 8540807 | 7253 | 9163312 | 9646 | 6973 | 76074 | 37801 | 13085 | 30 |
| 31 | 5839 | 8852440 | 8665 | 9495176 | 9832692 | 81997 | 43942 | 19460 | 29 |
| 32 | 8550873 | 7630 | 9174020 | 9500709 | 8415 | 87923 | 50087 | 25840 | 28 |
| 33 | 5910 | 8862822 | 9379 | 6245 | 9844141 | 93853 | 56235 | 32223 | 27 |
| 34 | 8560950 | 8017 | 9184740 | 9511784 | 9871 | 99786 | 62388 | 38610 | 26 |
| 35 | 5992 | 8873215 | 9190104 | 7326 | 9855603 | 1·0205723 | 68544 | 45002 | 25 |
| 36 | 8571037 | 8415 | 5471 | 9522871 | 9861339 | 11664 | 74704 | 51397 | 24 |
| 37 | 6084 | 8883619 | 9200841 | 8420 | 7079 | 17608 | 80867 | 57797 | 23 |
| 38 | 8581133 | 8825 | 6214 | 9533971 | 9872821 | 23555 | 87035 | 64201 | 22 |
| 39 | 6185 | 8894033 | 9211590 | 9526 | 8567 | 29506 | 93206 | 70609 | 21 |
| 40 | 8591240 | 9244 | 6969 | 9545083 | 9884316 | 35461 | 99381 | 77020 | 20 |
| 41 | 6297 | 8904458 | 9222350 | 9550644 | 9890069 | 41419 | 1·0605560 | 83436 | 19 |
| 42 | 8601357 | 9675 | 7734 | 6208 | 5825 | 47381 | 11742 | 89857 | 18 |
| 43 | 6419 | 8914894 | 9233122 | 9561774 | 9901584 | 53346 | 17929 | 96281 | 17 |
| 44 | 8611484 | 8920116 | 8512 | 7344 | 7346 | 59315 | 24119 | 1·1002709 | 16 |
| 45 | 6551 | 5341 | 9243905 | 9572917 | 9913112 | 65287 | 30313 | 09141 | 15 |
| 46 | 8621621 | 8930569 | 9301 | 8494 | 8881 | 71263 | 36511 | 15578 | 14 |
| 47 | 6694 | 5799 | 9254700 | 9584073 | 9924654 | 77243 | 42713 | 22019 | 13 |
| 48 | 8631768 | 8941032 | 9260102 | 9655 | 9930429 | 83226 | 48918 | 28463 | 12 |
| 49 | 6846 | 6268 | 5506 | 9595241 | 6208 | 89212 | 55128 | 34912 | 11 |
| 50 | 8641926 | 8951506 | 9270914 | 9600829 | 9941991 | 95203 | 61341 | 41365 | 10 |
| 51 | 7009 | 6747 | 6324 | 6421 | 7777 | 1·0301196 | 67558 | 47823 | 9 |
| 52 | 8652094 | 8961991 | 9281738 | 9612016 | 9953566 | 07194 | 73779 | 54284 | 8 |
| 53 | 7181 | 7238 | 7154 | 7614 | 9358 | 13195 | 80004 | 60750 | 7 |
| 54 | 8662272 | 8972487 | 9292573 | 9623215 | 9965154 | 19199 | 86233 | 67219 | 6 |
| 55 | 7365 | 7739 | 7996 | 8819 | 9970953 | 25208 | 92466 | 73693 | 5 |
| 56 | 8672460 | 8982994 | 9303421 | 9634427 | 6756 | 31220 | 98702 | 80171 | 4 |
| 57 | 7558 | 8251 | 8849 | 9640037 | 9982562 | 37235 | 1·0704943 | 86653 | 3 |
| 58 | 8682659 | 8993512 | 9314280 | 5651 | 8371 | 43254 | 11187 | 93140 | 2 |
| 59 | 7762 | 8775 | 9714 | 9651268 | 9994184 | 49277 | 17435 | 99630 | 1 |
| 60 | 869 867 | 9004040 | 9325151 | 6888 | 1·0000000 | 55303 | 23687 | 1·1106125 | 0 |
| ′ | 49° | 48° | 47° | 46° | 45° | 44° | 43° | 42° | |

| ′ | 48° | 49° | 50° | 51° | 52° | 53° | 54° | 55° | ′ |
|---|---|---|---|---|---|---|---|---|---|
| 0 | 743 1448 | 754 7096 | 766 0444 | 777 1460 | 788 0108 | 798 6355 | 809 0170 | 819 1520 | 60 |
| 1 | 3394 | 9004 | 2314 | 3290 | 1898 | 8105 | 1879 | 3189 | 59 |
| 2 | 5340 | 755 0911 | 4183 | 5120 | 3688 | 9855 | 3588 | 4856 | 58 |
| 3 | 7285 | 2818 | 6051 | 6949 | 5477 | 799 1604 | 5296 | 6523 | 57 |
| 4 | 9229 | 4724 | 7918 | 8777 | 7266 | 3352 | 7004 | 8189 | 56 |
| 5 | 744 1173 | 6630 | 9785 | 778 0604 | 9054 | 5100 | 8710 | 9854 | 55 |
| 6 | 3115 | 8535 | 767 1652 | 2431 | 789 0841 | 6847 | 810 0416 | 820 1519 | 54 |
| 7 | 5058 | 756 0439 | 3517 | 4258 | 2627 | 8593 | 2122 | 3183 | 53 |
| 8 | 6999 | 2342 | 5382 | 6084 | 4413 | 800 0338 | 3826 | 4846 | 52 |
| 9 | 8941 | 4246 | 7246 | 7909 | 6198 | 2083 | 5530 | 6509 | 51 |
| 10 | 745 0881 | 6148 | 9110 | 9733 | 7983 | 3827 | 7234 | 8170 | 50 |
| 11 | 2821 | 8050 | 768 0973 | 779 1557 | 9767 | 5571 | 8936 | 9832 | 49 |
| 12 | 4760 | 9951 | 2835 | 3380 | 790 1550 | 7314 | 811 0638 | 821 1492 | 48 |
| 13 | 6699 | 757 1851 | 4697 | 5202 | 3333 | 9056 | 2339 | 3152 | 47 |
| 14 | 8636 | 6751 | 6558 | 7024 | 5115 | 801 0797 | 4040 | 4811 | 46 |
| 15 | 746 0574 | 5650 | 8418 | 8845 | 6896 | 2538 | 5740 | 6469 | 45 |
| 16 | 2510 | 7548 | 769 0278 | 780 0665 | 8676 | 4278 | 7439 | 8127 | 44 |
| 17 | 4446 | 9446 | 2137 | 2485 | 791 0456 | 6018 | 9137 | 9784 | 43 |
| 18 | 6382 | 758 1343 | 3996 | 4304 | 2235 | 7756 | 812 0835 | 822 1440 | 42 |
| 19 | 8317 | 3240 | 5853 | 6123 | 4014 | 9495 | 2532 | 3096 | 41 |
| 20 | 747 0251 | 5136 | 7710 | 7940 | 5792 | 802 1232 | 4229 | 4751 | 40 |
| 21 | 2184 | 7031 | 9567 | 9757 | 7569 | 2969 | 5925 | 6405 | 39 |
| 22 | 4117 | 8926 | 770 1423 | 781 1574 | 9345 | 4705 | 7620 | 8059 | 38 |
| 23 | 6049 | 759 0820 | 3278 | 3390 | 792 1121 | 6440 | 9314 | 9712 | 37 |
| 24 | 7981 | 2713 | 5132 | 5205 | 2896 | 8175 | 813 1008 | 823 1364 | 36 |
| 25 | 9912 | 4606 | 6986 | 7019 | 4671 | 9909 | 2701 | 3015 | 35 |
| 26 | 748 1842 | 6498 | 8840 | 8833 | 6445 | 803 1642 | 4393 | 4666 | 34 |
| 27 | 3772 | 8389 | 771 0692 | 782 0646 | 8218 | 3375 | 6084 | 6316 | 33 |
| 28 | 5701 | 760 0280 | 2544 | 2459 | 9990 | 5107 | 7775 | 7965 | 32 |
| 29 | 7629 | 2170 | 4395 | 4270 | 793 1762 | 6838 | 9466 | 9614 | 31 |
| 30 | 9557 | 4060 | 6246 | 6082 | 3533 | 8569 | 814 1155 | 824 1262 | 30 |
| 31 | 749 1484 | 5949 | 8096 | 7892 | 5304 | 804 0299 | 2844 | 2909 | 29 |
| 32 | 3411 | 7837 | 9945 | 9702 | 7074 | 2028 | 4532 | 4556 | 28 |
| 33 | 5337 | 9724 | 772 1794 | 783 1511 | 8843 | 3756 | 6220 | 6202 | 27 |
| 34 | 7262 | 761 1611 | 3642 | 3320 | 794 0611 | 5484 | 7906 | 7847 | 26 |
| 35 | 9187 | 3497 | 5489 | 5127 | 2379 | 7211 | 9593 | 9491 | 25 |
| 36 | 750 1111 | 5383 | 7336 | 6935 | 4146 | 8938 | 815 1278 | 825 1135 | 24 |
| 37 | 3034 | 7268 | 9182 | 8741 | 5913 | 805 0664 | 2963 | 2778 | 23 |
| 38 | 4957 | 9152 | 773 1027 | 784 0547 | 7678 | 2389 | 4647 | 4420 | 22 |
| 39 | 6879 | 762 1036 | 2872 | 2352 | 9444 | 4113 | 6330 | 6062 | 21 |
| 40 | 8800 | 2919 | 4716 | 4157 | 795 1208 | 5837 | 8013 | 7703 | 20 |
| 41 | 751 0721 | 4802 | 6559 | 5961 | 2972 | 7560 | 9695 | 9343 | 19 |
| 42 | 2641 | 6683 | 8402 | 7764 | 4735 | 9283 | 816 1376 | 826 0983 | 18 |
| 43 | 4561 | 8564 | 774 0244 | 9566 | 6497 | 806 1005 | 3056 | 2622 | 17 |
| 44 | 6480 | 763 0445 | 2086 | 785 1368 | 8259 | 2726 | 4736 | 4260 | 16 |
| 45 | 8398 | 2325 | 3926 | 3169 | 796 0020 | 4446 | 6416 | 5897 | 15 |
| 46 | 752 0316 | 4204 | 5767 | 4970 | 1780 | 6166 | 8094 | 7534 | 14 |
| 47 | 2233 | 6082 | 7606 | 6770 | 3540 | 7885 | 9772 | 9170 | 13 |
| 48 | 4149 | 7960 | 9445 | 8569 | 5299 | 9603 | 817 1449 | 727 0806 | 12 |
| 49 | 6065 | 9838 | 775 1283 | 786 0367 | 7058 | 807 1321 | 3125 | 2440 | 11 |
| 50 | 7980 | 764 1714 | 3121 | 2165 | 8815 | 3038 | 4801 | 4074 | 10 |
| 51 | 9894 | 3590 | 4957 | 3963 | 797 0572 | 4754 | 6476 | 5708 | 9 |
| 52 | 753 1808 | 5465 | 6794 | 5759 | 2329 | 6470 | 8151 | 7340 | 8 |
| 53 | 3721 | 7340 | 8629 | 7555 | 4084 | 8185 | 9824 | 8972 | 7 |
| 54 | 5634 | 9214 | 776 0464 | 9350 | 5839 | 9899 | 818 1497 | 828 0603 | 6 |
| 55 | 7546 | 765 1087 | 2298 | 787 1145 | 7594 | 808 1612 | 3169 | 2234 | 5 |
| 56 | 9457 | 2960 | 4132 | 2939 | 9347 | 3325 | 4841 | 3864 | 4 |
| 57 | 754 1368 | 4832 | 5965 | 4732 | 798 1100 | 5037 | 6512 | 5493 | 3 |
| 58 | 3278 | 6704 | 7797 | 6524 | 2853 | 6749 | 8182 | 7121 | 2 |
| 59 | 5187 | 8574 | 9629 | 8316 | 4604 | 8460 | 9852 | 8749 | 1 |
| 60 | 7096 | 766 0444 | 777 1460 | 788 0108 | 6355 | 809 0170 | 819 1520 | 829 0376 | 0 |
| ′ | 41° | 40° | 39° | 38° | 37° | 36° | 35° | 34° | ′ |

NAT. COSINE.

| ′ | 48° | 49° | 50° | 51° | 52° | 53° | 54° | 55° | ′ |
|---|---|---|---|---|---|---|---|---|---|
| 0 | 1·1106125 | 1·1503684 | 1·1917536 | 1·2348972 | 1·2799416 | 1·3270448 | 1·3763819 | 1·4281480 | 60 |
| 1 | 12624 | 10445 | 24579 | 56319 | 1·2807094 | 78483 | 72242 | 90326 | 59 |
| 2 | 19127 | 17210 | 31626 | 63672 | 14776 | 86524 | 80672 | 99178 | 58 |
| 3 | 25635 | 23979 | 38679 | 71030 | 22465 | 94571 | 89108 | 1·4308039 | 57 |
| 4 | 32146 | 30754 | 45736 | 78393 | 30160 | 1·3302624 | 97551 | 16906 | 56 |
| 5 | 38662 | 37532 | 52799 | 85762 | 37860 | 10684 | 1·3806001 | 25781 | 55 |
| 6 | 45182 | 44316 | 59866 | 93136 | 45566 | 18750 | 14458 | 34664 | 54 |
| 7 | 51706 | 51104 | 66938 | 1·2400515 | 53277 | 26822 | 22922 | 43554 | 53 |
| 8 | 58235 | 57896 | 74015 | 07900 | 60995 | 34900 | 31392 | 52451 | 52 |
| 9 | 64768 | 64693 | 81097 | 15290 | 68718 | 42984 | 39869 | 61356 | 51 |
| 10 | 71305 | 71495 | 88184 | 22685 | 76447 | 51075 | 48353 | 70268 | 50 |
| 11 | 77846 | 78301 | 95276 | 30086 | 84182 | 59172 | 56844 | 79187 | 49 |
| 12 | 84391 | 85112 | 1·2002373 | 37492 | 91922 | 67276 | 65342 | 88114 | 48 |
| 13 | 90941 | 91927 | 09475 | 44903 | 99669 | 75386 | 73847 | 97049 | 47 |
| 14 | 97495 | 98747 | 16581 | 52320 | 1·2907421 | 83502 | 82358 | 1·4405991 | 46 |
| 15 | 1·1204053 | 1·1605571 | 23693 | 59742 | 15179 | 91624 | 90876 | 14940 | 45 |
| 16 | 10616 | 12400 | 30810 | 67169 | 22013 | 99753 | 99401 | 23897 | 44 |
| 17 | 17183 | 19234 | 37932 | 74602 | 30713 | 1·3407888 | 1·3907934 | 32862 | 43 |
| 18 | 23754 | 26073 | 45058 | 82040 | 38488 | 16029 | 16473 | 41834 | 42 |
| 19 | 30329 | 32916 | 52190 | 89484 | 46270 | 24177 | 25019 | 50814 | 41 |
| 20 | 36909 | 39763 | 59327 | 96933 | 54057 | 32331 | 33571 | 59801 | 40 |
| 21 | 43493 | 46615 | 66468 | 1·2504388 | 61850 | 40492 | 42131 | 68796 | 39 |
| 22 | 50081 | 53472 | 73615 | 11848 | 69649 | 48658 | 50698 | 77798 | 38 |
| 23 | 56674 | 60334 | 80767 | 19313 | 77454 | 56832 | 59272 | 86808 | 37 |
| 24 | 63271 | 67200 | 87924 | 26784 | 85265 | 65011 | 67852 | 95825 | 36 |
| 25 | 69872 | 74071 | 95065 | 34260 | 93081 | 73198 | 76440 | 1·4504850 | 35 |
| 26 | 76478 | 80947 | 1·2102252 | 41742 | 1·3000904 | 81390 | 85034 | 13883 | 34 |
| 27 | 83088 | 87827 | 09424 | 49229 | 08733 | 89589 | 93636 | 22923 | 33 |
| 28 | 89702 | 94712 | 16601 | 56721 | 16567 | 97794 | 1·4002245 | 31971 | 32 |
| 29 | 96321 | 1·1701601 | 23783 | 64219 | 24407 | 1·3506006 | 10860 | 41027 | 31 |
| 30 | 1·1302944 | 08496 | 30970 | 71723 | 32254 | 14224 | 19483 | 50090 | 30 |
| 31 | 09571 | 15395 | 38162 | 79232 | 40106 | 22449 | 28113 | 59161 | 29 |
| 32 | 16203 | 22298 | 45359 | 86747 | 47964 | 30680 | 36749 | 68240 | 28 |
| 33 | 22839 | 29207 | 52562 | 94267 | 55828 | 38918 | 45393 | 77326 | 27 |
| 34 | 29479 | 36120 | 59769 | 1·2601792 | 63699 | 47162 | 54044 | 86420 | 26 |
| 35 | 36124 | 43038 | 66982 | 09323 | 71575 | 55413 | 62702 | 95522 | 25 |
| 36 | 42773 | 49960 | 74199 | 16860 | 79457 | 63670 | 71367 | 1·4604632 | 24 |
| 37 | 49427 | 56888 | 81422 | 24402 | 87345 | 71934 | 80039 | 13749 | 23 |
| 38 | 56085 | 63820 | 88650 | 31950 | 95239 | 80204 | 88718 | 22874 | 22 |
| 39 | 62747 | 70756 | 95883 | 39503 | 1·3103140 | 88481 | 97405 | 32007 | 21 |
| 40 | 69414 | 77698 | ·2203121 | 47062 | 11046 | 96764 | 1·4106098 | 41147 | 20 |
| 41 | 76086 | 84644 | 10364 | 54626 | 18958 | 1·3605054 | 14799 | 50296 | 19 |
| 42 | 82761 | 91595 | 17613 | 62196 | 26876 | 13350 | 23506 | 59452 | 18 |
| 43 | 89441 | 98551 | 24866 | 69772 | 34801 | 21653 | 32221 | 68616 | 17 |
| 44 | 96126 | 1·1805512 | 32125 | 77353 | 42731 | 29963 | 40943 | 77788 | 16 |
| 45 | 1·1402815 | 12477 | 39389 | 84940 | 50668 | 38279 | 49673 | 86967 | 15 |
| 46 | 09508 | 19447 | 46658 | 92532 | 58610 | 46602 | 58409 | 96155 | 14 |
| 47 | 16206 | 26422 | 53932 | 1·2700130 | 66559 | 54931 | 67153 | 1·4705350 | 13 |
| 48 | 22908 | 33402 | 61211 | 07733 | 74513 | 63267 | 75904 | 14553 | 12 |
| 49 | 29615 | 40387 | 68496 | 15342 | 82474 | 71610 | 84662 | 23764 | 11 |
| 50 | 36326 | 47376 | 75786 | 22957 | 90441 | 79959 | 93427 | 32983 | 10 |
| 51 | 43041 | 54370 | 83081 | 30578 | 98414 | 88315 | 1·4202200 | 42210 | 9 |
| 52 | 49762 | 61369 | 90381 | 38204 | 1·3206393 | 96678 | 10979 | 51445 | 8 |
| 53 | 56486 | 68373 | 97687 | 45835 | 14379 | 1·3705047 | 19766 | 60688 | 7 |
| 54 | 63215 | 75382 | 1·2304997 | 53473 | 22370 | 13423 | 28561 | 69938 | 6 |
| 55 | 69949 | 82395 | 12313 | 61116 | 30368 | 21806 | 37362 | 79197 | 5 |
| 56 | 76687 | 89414 | 19634 | 68765 | 38371 | 30195 | 46171 | 88463 | 4 |
| 57 | 83429 | 96437 | 26961 | 76419 | 46381 | 38591 | 54988 | 97738 | 3 |
| 58 | 90176 | 1·1903465 | 34292 | 84079 | 54397 | 46994 | 63811 | 1·4807021 | 2 |
| 59 | 96928 | 10498 | 41629 | 91745 | 62420 | 55403 | 72642 | 16311 | 1 |
| 60 | 1·1503684 | 17536 | 48972 | 99416 | 70448 | 63819 | 81480 | 25610 | 0 |
| ′ | 41° | 40° | 39° | 38° | 37° | 36° | 35° | 34° | ′ |

NAT. COTAN.

| ′ | 56° | 57° | 58° | 59° | 60° | 61° | 62° | 63° | ′ |
|---|---|---|---|---|---|---|---|---|---|
| 0 | 829 0376 | 638 6706 | 848 0481 | 857 1673 | 866 0254 | 874 6197 | 982 9476 | 891 0065 | 60 |
| 1 | 2002 | 8290 | 2022 | 3171 | 1708 | 7607 | 883 0841 | 1385 | 59 |
| 2 | 3628 | 9873 | 3562 | 4668 | 3161 | 9016 | 2206 | 2705 | 58 |
| 3 | 5252 | 839 1455 | 5102 | 6164 | 4614 | 875 0425 | 3569 | 4024 | 57 |
| 4 | 6877 | 3037 | 6641 | 7660 | 6066 | 1832 | 4933 | 5342 | 56 |
| 5 | 8500 | 4618 | 8179 | 9155 | 7517 | 3239 | 6295 | 6659 | 55 |
| 6 | 830 0123 | 6199 | 9717 | 858 0649 | 8967 | 4645 | 7656 | 7975 | 54 |
| 7 | 1745 | 7778 | 849 1254 | 2143 | 867 0417 | 6051 | 9017 | 9291 | 53 |
| 8 | 3366 | 9357 | 2790 | 3635 | 1866 | 7455 | 884 0377 | 892 0606 | 52 |
| 9 | 4987 | 840 0936 | 4325 | 5127 | 3314 | 8859 | 1736 | 1920 | 51 |
| 10 | 6607 | 2513 | 5860 | 6619 | 4762 | 876 0263 | 3095 | 3234 | 50 |
| 11 | 8226 | 4090 | 7394 | 8109 | 6209 | 1665 | 4453 | 4546 | 49 |
| 12 | 9845 | 5666 | 6927 | 9599 | 7655 | 3067 | 5810 | 5858 | 48 |
| 13 | 831 1463 | 7241 | 850 0459 | 859 1088 | 9100 | 4468 | 7166 | 7169 | 47 |
| 14 | 3080 | 8816 | 1991 | 2576 | 868 0544 | 5868 | 8522 | 8480 | 46 |
| 15 | 4696 | 841 0390 | 3522 | 4064 | 1988 | 7268 | 9876 | 9789 | 45 |
| 16 | 6312 | 1963 | 5053 | 5551 | 3431 | 8666 | 885 1230 | 893 1098 | 44 |
| 17 | 7927 | 3536 | 6582 | 7037 | 4874 | 877 0064 | 2584 | 2406 | 43 |
| 18 | 9541 | 5108 | 8111 | 8523 | 6315 | 1462 | 3936 | 3714 | 42 |
| 19 | 832 1155 | 6679 | 9639 | 860 0007 | 7756 | 2858 | 5288 | 5021 | 41 |
| 20 | 2768 | 8249 | 851 1167 | 1491 | 9196 | 4254 | 6639 | 6326 | 40 |
| 21 | 4380 | 9819 | 2693 | 2975 | 869 0636 | 5649 | 7989 | 7632 | 39 |
| 22 | 5991 | 842 1388 | 4219 | 4457 | 2074 | 7043 | 9339 | 8936 | 38 |
| 23 | 7602 | 2956 | 5745 | 5939 | 3512 | 8437 | 886 0688 | 894 0240 | 37 |
| 24 | 9212 | 4524 | 7269 | 7420 | 4949 | 9830 | 2036 | 1542 | 36 |
| 25 | 833 0822 | 6091 | 8793 | 8901 | 6386 | 878 1222 | 3383 | 2844 | 35 |
| 26 | 2430 | 7657 | 852 0316 | 861 0380 | 7821 | 2613 | 4730 | 4146 | 34 |
| 27 | 4038 | 9222 | 1839 | 1859 | 9256 | 4004 | 6075 | 5446 | 33 |
| 28 | 5646 | 843 0787 | 3360 | 3337 | 870 0691 | 5394 | 7420 | 6746 | 32 |
| 29 | 7252 | 2351 | 4881 | 4815 | 2124 | 6783 | 8765 | 8045 | 31 |
| 30 | 8858 | 3914 | 6402 | 6292 | 3557 | 8171 | 887 0108 | 9344 | 30 |
| 31 | 834 0463 | 5477 | 7921 | 7768 | 4989 | 9559 | 1451 | 895 0641 | 29 |
| 32 | 2068 | 7039 | 9440 | 9243 | 6420 | 879 0946 | 2793 | 1938 | 28 |
| 33 | 3672 | 8600 | 853 0958 | 862 0717 | 7851 | 2332 | 4134 | 3234 | 27 |
| 34 | 5275 | 844 0161 | 2475 | 2191 | 9281 | 3717 | 5475 | 4529 | 26 |
| 35 | 6877 | 1720 | 3992 | 5664 | 871 0710 | 5102 | 6815 | 5824 | 25 |
| 36 | 8479 | 3279 | 5508 | 5137 | 2138 | 6486 | 8154 | 7118 | 24 |
| 37 | 835 0080 | 4838 | 7023 | 6608 | 3566 | 7869 | 9492 | 8411 | 23 |
| 38 | 1680 | 6395 | 8538 | 8079 | 4993 | 9251 | 888 0830 | 9703 | 22 |
| 39 | 3279 | 7952 | 854 0051 | 9549 | 6419 | 880 0633 | 2166 | 896 0994 | 21 |
| 40 | 4878 | 9508 | 1564 | 863 1019 | 7844 | 2014 | 3503 | 2285 | 20 |
| 41 | 6476 | 845 1064 | 3077 | 2488 | 9269 | 3394 | 4838 | 3575 | 19 |
| 42 | 8074 | 2618 | 4588 | 3956 | 872 0693 | 4774 | 6172 | 4864 | 18 |
| 43 | 9670 | 4172 | 6099 | 5423 | 2116 | 6152 | 7506 | 6153 | 17 |
| 44 | 836 1266 | 5726 | 7609 | 6889 | 3538 | 7530 | 8839 | 7440 | 16 |
| 45 | 2862 | 7278 | 9119 | 8355 | 4960 | 8907 | 889 0171 | 8727 | 15 |
| 46 | 4456 | 8830 | 855 0627 | 9820 | 6381 | 881 0284 | 1503 | 897 0014 | 14 |
| 47 | 6050 | 846 0381 | 2135 | 864 1284 | 7801 | 1660 | 2834 | 1299 | 13 |
| 48 | 7643 | 1932 | 3643 | 2748 | 9221 | 3035 | 4164 | 2584 | 12 |
| 49 | 9236 | 3481 | 5149 | 4211 | 873 0640 | 4409 | 5493 | 3868 | 11 |
| 50 | 837 0827 | 5030 | 6655 | 5673 | 2058 | 5782 | 6822 | 5151 | 10 |
| 51 | 2418 | 6579 | 8160 | 7134 | 3475 | 7155 | 8149 | 6433 | 9 |
| 52 | 4009 | 8126 | 9664 | 8595 | 4891 | 8527 | 9476 | 7715 | 8 |
| 53 | 5598 | 9673 | 856 1168 | 865 0055 | 6307 | 9898 | 890 0803 | 8996 | 7 |
| 54 | 7187 | 847 1219 | 2671 | 1514 | 7722 | 882 1269 | 2128 | 898 0276 | 6 |
| 55 | 8775 | 2765 | 4173 | 2973 | 9137 | 2638 | 3453 | 1555 | 5 |
| 56 | 838 0363 | 4309 | 5674 | 4430 | 874 0550 | 4007 | 4777 | 2834 | 4 |
| 57 | 1950 | 5853 | 7175 | 5887 | 1963 | 5376 | 6100 | 4112 | 3 |
| 58 | 3536 | 7397 | 8675 | 7344 | 3375 | 6743 | 7423 | 5389 | 2 |
| 59 | 5121 | 8939 | 857 0174 | 8799 | 4786 | 8110 | 8744 | 6665 | 1 |
| 60 | 6706 | 848 0481 | 1673 | 866 0254 | 6197 | 9476 | 891 0065 | 7940 | 0 |
| ′ | 33° | 32° | 31° | 30° | 29° | 28° | 27° | 26° | ′ |

| ′ | 56° | 57° | 58° | 59° | 60° | 61° | 62° | 63° | ′ |
|---|---|---|---|---|---|---|---|---|---|
| 0 | 1·4825610 | 1·5 98650 | 1·6003345 | 1·6642795 | 1·7320508 | 1·8040478 | 1·8807265 | 1·9626105 | 60 |
| 1 | 34916 | 1·5408460 | 13709 | 53766 | 32149 | 52860 | 20470 | 40227 | 59 |
| 2 | 44231 | 18280 | 24082 | 64748 | 43803 | 65256 | 33690 | 54364 | 58 |
| 3 | 53554 | 28108 | 34465 | 75741 | 55468 | 77664 | 46924 | 68518 | 57 |
| 4 | 62884 | 37946 | 44858 | 86744 | 67144 | 90086 | 60172 | 82688 | 56 |
| 5 | 72223 | 47792 | 55260 | 97758 | 78833 | 1·8102521 | 73436 | 96874 | 55 |
| 6 | 81570 | 57647 | 65672 | 1·6708782 | 90533 | 14969 | 86713 | 1·9711077 | 54 |
| 7 | 90925 | 67510 | 76094 | 19818 | 1·7402245 | 27430 | 1·8900006 | 25296 | 53 |
| 8 | 1·4900288 | 77383 | 86525 | 30864 | 13969 | 39904 | 13313 | 39531 | 52 |
| 9 | 09659 | 87264 | 96966 | 41921 | 25705 | 52391 | 26635 | 53782 | 51 |
| 10 | 19039 | 97155 | 1·6107417 | 52988 | 37453 | 64892 | 39971 | 68050 | 50 |
| 11 | 28426 | 1·5507054 | 17878 | 64067 | 49213 | 77405 | 53322 | 82334 | 49 |
| 12 | 37822 | 16963 | 28349 | 75156 | 60984 | 89932 | 66688 | 96635 | 48 |
| 13 | 47225 | 26880 | 38829 | 86256 | 72768 | 1·8202473 | 80068 | 1·9810952 | 47 |
| 14 | 56637 | 36806 | 49320 | 97367 | 84564 | 15026 | 93464 | 25286 | 46 |
| 15 | 66058 | 46741 | 59820 | 1·6808489 | 96371 | 27593 | 1·9006874 | 39636 | 45 |
| 16 | 75486 | 56685 | 70330 | 19621 | 1·7508191 | 40173 | 20299 | 54003 | 44 |
| 17 | 84923 | 66639 | 80850 | 30765 | 20023 | 52767 | 33738 | 68387 | 43 |
| 18 | 94367 | 76601 | 91380 | 41919 | 31866 | 65374 | 47193 | 82787 | 42 |
| 19 | 1·5003821 | 86572 | 1·6201920 | 53085 | 43722 | 77994 | 60663 | 97204 | 41 |
| 20 | 13282 | 96552 | 12469 | 64261 | 55590 | 90628 | 74147 | 1·9911637 | 40 |
| 21 | 22751 | 1·5606542 | 23029 | 75449 | 67470 | 1·8303275 | 87647 | 26087 | 39 |
| 22 | 32229 | 16540 | 33599 | 86647 | 79362 | 15936 | 1·9101162 | 40554 | 38 |
| 23 | 41716 | 26548 | 44178 | 97856 | 91267 | 28610 | 14691 | 55038 | 37 |
| 24 | 51210 | 36564 | 54768 | 1·6909077 | 1·7603183 | 41297 | 28236 | 69539 | 36 |
| 25 | 60713 | 46590 | 65368 | 20308 | 15112 | 53999 | 41795 | 84056 | 35 |
| 26 | 70224 | 56625 | 75977 | 31550 | 27053 | 66713 | 55370 | 98590 | 34 |
| 27 | 79743 | 66669 | 86597 | 42804 | 39007 | 79442 | 68960 | 2·0013142 | 33 |
| 28 | 89271 | 76722 | 97227 | 54069 | 50972 | 92184 | 82565 | 27710 | 32 |
| 29 | 98807 | 86784 | 1·6307867 | 65344 | 62950 | 1·8404940 | 96186 | 42295 | 31 |
| 30 | 1·5108352 | 96856 | 18517 | 76631 | 74940 | 17709 | 1·9209821 | 56897 | 30 |
| 31 | 17905 | 1·5706936 | 29177 | 87929 | 86943 | 30492 | 23472 | 71516 | 29 |
| 32 | 27466 | 17026 | 39847 | 99238 | 98958 | 43289 | 37138 | 86153 | 28 |
| 33 | 37036 | 27126 | 50528 | 1·7010559 | 1·7710985 | 56099 | 50819 | 2·0100806 | 27 |
| 34 | 46614 | 37234 | 61218 | 21890 | 23024 | 68923 | 64516 | 15477 | 26 |
| 35 | 56201 | 47352 | 71919 | 33233 | 35076 | 81761 | 78228 | 30164 | 25 |
| 36 | 65796 | 57479 | 82630 | 44587 | 47141 | 94613 | 91956 | 44869 | 24 |
| 37 | 75400 | 67615 | 93351 | 55953 | 59218 | 1·8507479 | 1·9305699 | 59592 | 23 |
| 38 | 85012 | 77760 | 1·6404082 | 67329 | 71307 | 20358 | 19457 | 74331 | 22 |
| 39 | 94632 | 87915 | 14824 | 78717 | 83409 | 33252 | 33231 | 89088 | 21 |
| 40 | 1·5204261 | 98079 | 25576 | 90116 | 95524 | 46159 | 47020 | 2·0203862 | 20 |
| 41 | 13899 | 1·5808253 | 36338 | 1·7101527 | 1·7807651 | 59080 | 60825 | 18654 | 19 |
| 42 | 23545 | 18436 | 47111 | 12949 | 19790 | 72015 | 74645 | 33462 | 18 |
| 43 | 33200 | 28628 | 57893 | 24382 | 31943 | 84965 | 88481 | 48289 | 17 |
| 44 | 42863 | 38830 | 68687 | 35827 | 44107 | 97928 | 1·9402333 | 63133 | 16 |
| 45 | 52535 | 49041 | 79490 | 47283 | 56285 | 1·8610905 | 16200 | 77994 | 15 |
| 46 | 62215 | 59261 | 90304 | 58751 | 68475 | 23896 | 30083 | 92873 | 14 |
| 47 | 71904 | 69491 | 1·6501128 | 70230 | 80678 | 36902 | 43981 | 2·0307769 | 13 |
| 48 | 81602 | 79731 | 11963 | 81720 | 92893 | 49921 | 57896 | 22683 | 12 |
| 49 | 91308 | 89979 | 22808 | 93222 | 1·7905121 | 62955 | 71826 | 37615 | 11 |
| 50 | 1·5301023 | 1·5900238 | 33663 | 1·7204736 | 17362 | 76003 | 85772 | 52565 | 10 |
| 51 | 10746 | 10505 | 44529 | 16261 | 29616 | 89065 | 99733 | 67532 | 9 |
| 52 | 20479 | 20783 | 55405 | 27797 | 41883 | 1·8702141 | 1·9513711 | 82517 | 8 |
| 53 | 30219 | 31070 | 66292 | 39346 | 54162 | 15231 | 27704 | 97519 | 7 |
| 54 | 39969 | 41366 | 77189 | 50905 | 66454 | 28336 | 41713 | 2·0412540 | 6 |
| 55 | 49727 | 51672 | 88097 | 62477 | 78759 | 41455 | 55739 | 27578 | 5 |
| 56 | 59494 | 61987 | 99016 | 74060 | 91077 | 54588 | 69780 | 42634 | 4 |
| 57 | 69270 | 72312 | 1·6609945 | 85654 | 1·8003408 | 67736 | 83837 | 57708 | 3 |
| 58 | 79054 | 82647 | 20884 | 97260 | 15751 | 80898 | 97910 | 72800 | 2 |
| 59 | 88848 | 92991 | 31834 | 1·7308878 | 28108 | 94074 | 1·9612000 | 87910 | 1 |
| 60 | 98650 | 1·6003345 | 42795 | 20508 | 40478 | 1·8807265 | 26105 | 2·0503038 | 0 |
| ′ | 33° | 32° | 31° | 30° | 29° | 28° | 27° | 26° | ′ |

NAT. SINE.

| ′ | 64° | 65° | 66° | 67° | 68° | 69° | 70° | 71° | ′ |
|---|---|---|---|---|---|---|---|---|---|
| 0 | 898 7940 | 906 3078 | 913 5455 | 920 5049 | 927 1839 | 933 5804 | 939 6926 | 945 5186 | 60 |
| 1 | 9215 | 4307 | 6637 | 6185 | 2928 | 6846 | 7921 | 6132 | 59 |
| 2 | 899 0489 | 5535 | 7819 | 7320 | 4016 | 7888 | 8914 | 7078 | 58 |
| 3 | 1763 | 6762 | 9001 | 8455 | 5104 | 8928 | 9907 | 8023 | 57 |
| 4 | 3035 | 7989 | 914 0181 | 9589 | 6191 | 9968 | 940 0899 | 8968 | 56 |
| 5 | 4307 | 9215 | 1361 | 921 0722 | 7277 | 934 1007 | 1891 | 9911 | 55 |
| 6 | 5578 | 907 0440 | 2540 | 1854 | 8363 | 2045 | 2881 | 946 0854 | 54 |
| 7 | 6848 | 1665 | 3718 | 2986 | 9447 | 3082 | 3871 | 1795 | 53 |
| 8 | 8117 | 2888 | 4895 | 4116 | 928 0531 | 4119 | 4860 | 2736 | 52 |
| 9 | 9386 | 4111 | 6072 | 5246 | 1614 | 5154 | 5848 | 3677 | 51 |
| 10 | 900 0654 | 5333 | 7247 | 6375 | 2696 | 6189 | 6835 | 4616 | 50 |
| 11 | 1921 | 6554 | 8422 | 7504 | 3778 | 7223 | 7822 | 5555 | 49 |
| 12 | 3188 | 7775 | 9597 | 8632 | 4858 | 8257 | 8808 | 6493 | 48 |
| 13 | 4453 | 8995 | 915 0770 | 9758 | 5938 | 9289 | 9793 | 7430 | 47 |
| 14 | 5718 | 908 0214 | 1943 | 922 0884 | 7017 | 935 0321 | 941 0777 | 8366 | 46 |
| 15 | 6982 | 1432 | 3115 | 2010 | 8096 | 1352 | 1760 | 9301 | 45 |
| 16 | 8246 | 2649 | 4286 | 3134 | 9173 | 2382 | 2743 | 947 0236 | 44 |
| 17 | 9508 | 3866 | 5456 | 4258 | 929 0250 | 3412 | 3724 | 1170 | 43 |
| 18 | 901 0770 | 5082 | 6626 | 5381 | 1326 | 4440 | 4705 | 2103 | 42 |
| 19 | 2031 | 6297 | 7795 | 6503 | 2401 | 5468 | 5686 | 3035 | 41 |
| 20 | 3292 | 7511 | 8963 | 7624 | 3475 | 6495 | 6665 | 3966 | 40 |
| 21 | 4551 | 8725 | 916 0130 | 8745 | 4549 | 7521 | 7644 | 4897 | 39 |
| 22 | 5810 | 9938 | 1297 | 9865 | 5622 | 8547 | 8621 | 5827 | 38 |
| 23 | 7068 | 909 1150 | 2462 | 923 0984 | 6694 | 9571 | 9598 | 6756 | 37 |
| 24 | 8325 | 2361 | 3627 | 2102 | 7765 | 936 0595 | 942 0575 | 7684 | 36 |
| 25 | 9582 | 3572 | 4791 | 3220 | 8835 | 1618 | 1550 | 8612 | 35 |
| 26 | 902 0838 | 4781 | 5955 | 4336 | 9905 | 2641 | 2525 | 9538 | 34 |
| 27 | 2092 | 5990 | 7118 | 5452 | 930 0974 | 3662 | 3498 | 948 0464 | 33 |
| 28 | 3347 | 7199 | 8279 | 6567 | 2042 | 4683 | 4471 | 1389 | 32 |
| 29 | 4600 | 8406 | 9440 | 7682 | 3109 | 5703 | 5444 | 2313 | 31 |
| 30 | 5853 | 9613 | 917 0601 | 8795 | 4176 | 6722 | 6415 | 3237 | 30 |
| 31 | 7105 | 910 0819 | 1760 | 9908 | 5241 | 7740 | 7386 | 4159 | 29 |
| 32 | 8356 | 2024 | 2919 | 924 1020 | 6306 | 8758 | 8355 | 5081 | 28 |
| 33 | 9606 | 3228 | 4077 | 2131 | 7370 | 9774 | 9324 | 6002 | 27 |
| 34 | 903 0856 | 4432 | 5234 | 3242 | 8434 | 937 0790 | 943 0293 | 6922 | 26 |
| 35 | 2105 | 5635 | 6391 | 4351 | 9496 | 1806 | 1260 | 7842 | 25 |
| 36 | 3353 | 6837 | 7546 | 5460 | 931 0558 | 2820 | 2227 | 8760 | 24 |
| 37 | 4600 | 8038 | 8701 | 6568 | 1619 | 3833 | 3192 | 9678 | 23 |
| 38 | 5847 | 9236 | 9855 | 7676 | 2679 | 4846 | 4157 | 949 0595 | 22 |
| 39 | 7093 | 911 0438 | 918 1009 | 8782 | 3739 | 5858 | 5122 | 1511 | 21 |
| 40 | 8338 | 1637 | 2161 | 9888 | 4797 | 6869 | 6085 | 2426 | 20 |
| 41 | 9582 | 2835 | 3313 | 925 0993 | 5855 | 7880 | 7048 | 3341 | 19 |
| 42 | 904 0825 | 4033 | 4464 | 2097 | 6912 | 8889 | 8010 | 4255 | 18 |
| 43 | 2068 | 5229 | 5614 | 3201 | 7969 | 9898 | 8971 | 5168 | 17 |
| 44 | 3310 | 6425 | 6763 | 4303 | 9024 | 938 0906 | 9931 | 6080 | 16 |
| 45 | 4551 | 7620 | 7912 | 5405 | 932 0079 | 1913 | 944 0890 | 6991 | 15 |
| 46 | 5792 | 8815 | 9060 | 6506 | 1133 | 2920 | 1849 | 7902 | 14 |
| 47 | 7032 | 912 0008 | 919 0207 | 7606 | 2186 | 3925 | 2807 | 8812 | 13 |
| 48 | 8271 | 1201 | 1353 | 8706 | 3238 | 4930 | 3764 | 9721 | 12 |
| 49 | 9509 | 2393 | 2499 | 9805 | 4290 | 5934 | 4720 | 950 0629 | 11 |
| 50 | 905 0746 | 3584 | 3644 | 926 0902 | 5340 | 6938 | 5675 | 1536 | 10 |
| 51 | 1983 | 4775 | 4788 | 2000 | 6390 | 7940 | 6630 | 2443 | 9 |
| 52 | 3219 | 5965 | 5931 | 3096 | 7439 | 8942 | 7584 | 3348 | 8 |
| 53 | 4454 | 7154 | 7073 | 4192 | 8488 | 9943 | 8537 | 4253 | 7 |
| 54 | 5688 | 8342 | 8215 | 5286 | 9535 | 939 0943 | 9489 | 5157 | 6 |
| 55 | 6922 | 9529 | 9356 | 6380 | 933 0582 | 1942 | 945 0441 | 6061 | 5 |
| 56 | 8154 | 913 0716 | 920 0496 | 7474 | 1628 | 2940 | 1391 | 6963 | 4 |
| 57 | 9386 | 1902 | 1635 | 8566 | 2673 | 3938 | 2341 | 7865 | 3 |
| 58 | 906 0618 | 3087 | 2774 | 9658 | 3718 | 4935 | 3290 | 8766 | 2 |
| 59 | 1848 | 4271 | 3912 | 927 0748 | 4761 | 5931 | 4238 | 9666 | 1 |
| 60 | 3078 | 5455 | 5049 | 1839 | 5804 | 6926 | 5186 | 951 0565 | 0 |
| ′ | 25° | 24° | 23° | 22° | 21° | 20° | 19° | 18° | ′ |

NAT. COSINE.

| ′ | 64° | 65° | 66° | 67° | 68° | 69° | 70° | 71° | ′ |
|---|---|---|---|---|---|---|---|---|---|
| 0 | 2·0503038 | 2·1445069 | 2·2460368 | 2·3558524 | 2·4750869 | 2·6050891 | 2·7474774 | 2·9042109 | 60 |
| 1 | 18185 | 61366 | 77962 | 77590 | 71612 | 73558 | 99661 | 69576 | 59 |
| 2 | 33349 | 77683 | 95580 | 96683 | 92386 | 96259 | 2·7524588 | 97089 | 58 |
| 3 | 48531 | 94021 | 2·2513221 | 2·3615801 | 2·4813190 | 2·6118995 | 49554 | 2·9124649 | 57 |
| 4 | 63732 | 2·1510378 | 30885 | 34946 | 34023 | 41766 | 74561 | 52256 | 56 |
| 5 | 78950 | 26757 | 48572 | 54118 | 54887 | 64571 | 99608 | 79909 | 55 |
| 6 | 94187 | 43156 | 66283 | 73316 | 75781 | 87411 | 2·7624695 | 2·9207610 | 54 |
| 7 | 2·0609442 | 59575 | 84016 | 92540 | 96706 | 2·6210286 | 49822 | 35358 | 53 |
| 8 | 24716 | 76015 | 2·2601773 | 2·3711791 | 2·4917660 | 33196 | 74990 | 63152 | 52 |
| 9 | 40008 | 92476 | 19554 | 31068 | 38645 | 56141 | 2·7700199 | 90995 | 51 |
| 10 | 55318 | 2·1608958 | 37357 | 50372 | 59661 | 79121 | 25448 | 2·9318885 | 50 |
| 11 | 70646 | 25460 | 55184 | 69703 | 80707 | 2·6302136 | 50738 | 46822 | 49 |
| 12 | 85994 | 41983 | 73035 | 89060 | 2·5001784 | 25186 | 76069 | 74807 | 48 |
| 13 | 2·0701359 | 58527 | 90909 | 2·3808444 | 22891 | 48271 | 2·7801440 | 2·9402840 | 47 |
| 14 | 16743 | 75091 | 2·2708807 | 27855 | 44029 | 71392 | 26853 | 30921 | 46 |
| 15 | 32146 | 91677 | 26729 | 47293 | 65198 | 94549 | 52307 | 59050 | 45 |
| 16 | 47567 | 2·1708283 | 44674 | 66758 | 86398 | 2·6417741 | 77802 | 87227 | 44 |
| 17 | 63007 | 24911 | 62643 | 86250 | 2·5107629 | 40969 | 2·7903339 | 2·9515453 | 43 |
| 18 | 78465 | 41559 | 80636 | 2·3905769 | 28890 | 64232 | 28917 | 43727 | 42 |
| 19 | 93942 | 58229 | 98653 | 25316 | 50183 | 87531 | 54537 | 72050 | 41 |
| 20 | 2·0809438 | 74920 | 2·2816693 | 44889 | 71507 | 2·6510867 | 80198 | 2·9600422 | 40 |
| 21 | 24953 | 91631 | 34758 | 64490 | 92863 | 34238 | 2·8005901 | 28842 | 39 |
| 22 | 40487 | 2·1808364 | 52846 | 84118 | 2·5214249 | 57645 | 31646 | 57312 | 38 |
| 23 | 56039 | 25119 | 70959 | 2·4003774 | 35667 | 81089 | 57433 | 85831 | 37 |
| 24 | 71610 | 41894 | 89096 | 23457 | 57117 | 2·6604569 | 83263 | 2·9714399 | 36 |
| 25 | 87200 | 58691 | 2·2907257 | 43168 | 78598 | 28085 | 2·8109164 | 43016 | 35 |
| 26 | 2·0902809 | 75510 | 25442 | 62906 | 2·5300111 | 51638 | 35048 | 71683 | 34 |
| 27 | 18437 | 92349 | 43651 | 82672 | 21655 | 75227 | 61004 | 2·9800400 | 33 |
| 28 | 34085 | 2·1909210 | 61885 | 2·4102465 | 43231 | 98853 | 87003 | 29167 | 32 |
| 29 | 49751 | 26093 | 80143 | 22286 | 64839 | 2·6722516 | 2·8213045 | 57983 | 31 |
| 30 | 65436 | 42997 | 98425 | 42136 | 86479 | 46215 | 39129 | 86850 | 30 |
| 31 | 81140 | 59923 | 2·3016732 | 62013 | 2·5408151 | 69951 | 65256 | 2·9915766 | 29 |
| 32 | 96864 | 76871 | 35064 | 81918 | 29855 | 93725 | 91426 | 44734 | 28 |
| 33 | 2·1012607 | 93840 | 53420 | 2·4201851 | 51591 | 2·6817535 | 2·8317639 | 73751 | 27 |
| 34 | 28369 | 2·2010831 | 71801 | 21812 | 73359 | 41383 | 43896 | 3·0002820 | 26 |
| 35 | 44150 | 27843 | 90206 | 41801 | 95160 | 65267 | 70196 | 31939 | 25 |
| 36 | 59951 | 44878 | 2·3108637 | 61819 | 2·5516992 | 89190 | 96539 | 61109 | 24 |
| 37 | 75771 | 61934 | 27092 | 81864 | 38858 | 2·6913149 | 2·8422926 | 90330 | 23 |
| 38 | 91611 | 79012 | 45571 | 2·4301938 | 60756 | 37147 | 49356 | 3·0119603 | 22 |
| 39 | 2·1107470 | 96112 | 64076 | 22041 | 82686 | 61181 | 75831 | 48926 | 21 |
| 40 | 23348 | 2·2113234 | 82606 | 42172 | 2·5604649 | 85254 | 2·8502349 | 78301 | 20 |
| 41 | 39246 | 30379 | 2·3201160 | 62331 | 26645 | 2·7009364 | 28911 | 3·0207728 | 19 |
| 42 | 55164 | 47545 | 19740 | 82519 | 48674 | 33513 | 55517 | 37207 | 18 |
| 43 | 71101 | 64733 | 38345 | 2·4402736 | 70735 | 57699 | 82168 | 66737 | 17 |
| 44 | 87057 | 81944 | 56975 | 22982 | 92830 | 81923 | 2·8608863 | 96320 | 16 |
| 45 | 2·1203034 | 99177 | 75630 | 43256 | 2·5714957 | 2·7106186 | 35602 | 3·0325954 | 15 |
| 46 | 19030 | 2·2216432 | 94311 | 63559 | 37118 | 30487 | 62386 | 55641 | 14 |
| 47 | 35046 | 33709 | 2·3313017 | 83891 | 59312 | 54826 | 89215 | 85381 | 13 |
| 48 | 51082 | 51009 | 31748 | 2·4504252 | 81539 | 79204 | 2·8716088 | 3·0415173 | 12 |
| 49 | 67137 | 68331 | 50505 | 24642 | 2·5803800 | 2·7203620 | 43007 | 45018 | 11 |
| 50 | 83213 | 85676 | 69287 | 45061 | 26094 | 28076 | 69970 | 74915 | 10 |
| 51 | 99308 | 2·2303043 | 88095 | 65510 | 48421 | 52569 | 96979 | 3·0504866 | 9 |
| 52 | 2·1315423 | 20433 | 2·3406928 | 85987 | 70782 | 77102 | 2·8824033 | 34870 | 8 |
| 53 | 31559 | 37845 | 25787 | 2·4606494 | 93177 | 2·7301674 | 51132 | 64928 | 7 |
| 54 | 47714 | 55280 | 44672 | 27030 | 2·5915606 | 26284 | 78277 | 95038 | 6 |
| 55 | 63890 | 72738 | 63582 | 47596 | 38068 | 50934 | 2·8905467 | 3·0625203 | 5 |
| 56 | 80085 | 90218 | 82519 | 68191 | 60564 | 75623 | 32704 | 55421 | 4 |
| 57 | 96301 | 2·2407721 | 2·3501481 | 88816 | 83095 | 2·7400352 | 59986 | 85694 | 3 |
| 58 | 2·1412537 | 25247 | 20469 | 2·4709470 | 2·6005659 | 25120 | 87314 | 3·0716020 | 2 |
| 59 | 28793 | 42796 | 39483 | 30155 | 28258 | 49927 | 2·9014688 | 46400 | 1 |
| 60 | 45069 | 60368 | 58524 | 50869 | 50891 | 74774 | 42109 | 76835 | 0 |
| ′ | 25° | 24° | 23° | 22° | 21° | 20° | 19° | 18° | ′ |

| ′ | 72° | 73° | 74° | 75° | 76° | 77° | 78° | 79° | ′ |
|---|---|---|---|---|---|---|---|---|---|
| 0 | 951 0565 | 956 3048 | 961 2617 | 965 9258 | 970 2957 | 974 3701 | 978 1476 | 981 6272 | 60 |
| 1 | 1464 | 3898 | 3418 | 966 0011 | 3661 | 4355 | 2080 | 6826 | 59 |
| 2 | 2361 | 4747 | 4219 | 0762 | 4363 | 5008 | 2684 | 7380 | 58 |
| 3 | 3258 | 5595 | 5019 | 1513 | 5065 | 5660 | 3287 | 7933 | 57 |
| 4 | 4154 | 6443 | 5818 | 2263 | 5766 | 6311 | 3889 | 8485 | 56 |
| 5 | 5050 | 7290 | 6616 | 3012 | 6466 | 6962 | 4490 | 9037 | 55 |
| 6 | 5944 | 8136 | 7413 | 3761 | 7165 | 7612 | 5090 | 9587 | 54 |
| 7 | 6838 | 8981 | 8210 | 4508 | 7863 | 8261 | 5689 | 982 0137 | 53 |
| 8 | 7731 | 9825 | 9005 | 5255 | 8561 | 8909 | 6288 | 0686 | 52 |
| 9 | 8623 | 957 0669 | 9800 | 6001 | 9258 | 9556 | 6886 | 1234 | 51 |
| 10 | 9514 | 1512 | 962 0594 | 6746 | 9953 | 975 0203 | 7483 | 1781 | 50 |
| 11 | 952 0404 | 2354 | 1387 | 7490 | 971 0649 | 0849 | 8079 | 2327 | 49 |
| 12 | 1294 | 3195 | 2180 | 8234 | 1343 | 1494 | 8674 | 2873 | 48 |
| 13 | 2183 | 4035 | 2972 | 8977 | 2036 | 2138 | 9268 | 3417 | 47 |
| 14 | 3071 | 4875 | 3762 | 9718 | 2729 | 2781 | 9862 | 3961 | 46 |
| 15 | 3958 | 5714 | 4552 | 967 0459 | 3421 | 3423 | 979 0455 | 4504 | 45 |
| 16 | 4844 | 6552 | 5342 | 1200 | 4112 | 4065 | 1047 | 5046 | 44 |
| 17 | 5730 | 7389 | 6130 | 1939 | 4802 | 4706 | 1638 | 5587 | 43 |
| 18 | 6615 | 8225 | 6917 | 2678 | 5491 | 5345 | 2228 | 6128 | 42 |
| 19 | 7499 | 9060 | 7704 | 3415 | 6180 | 5985 | 2818 | 6668 | 41 |
| 20 | 8382 | 9895 | 8490 | 4152 | 6867 | 6623 | 3406 | 7206 | 40 |
| 21 | 9264 | 958 0729 | 9275 | 4888 | 7554 | 7260 | 3994 | 7744 | 39 |
| 22 | 953 0146 | 1562 | 963 0060 | 5624 | 8240 | 7897 | 4581 | 8282 | 38 |
| 23 | 1027 | 2394 | 0843 | 6358 | 8926 | 8533 | 5167 | 8818 | 37 |
| 24 | 1907 | 3226 | 1626 | 7092 | 9610 | 9168 | 5752 | 9353 | 36 |
| 25 | 2786 | 4056 | 2408 | 7825 | 972 0294 | 9802 | 6337 | 9888 | 35 |
| 26 | 3664 | 4886 | 3189 | 8557 | 0976 | 976 0435 | 6921 | 983 0422 | 34 |
| 27 | 4542 | 5715 | 3969 | 9288 | 1658 | 1068 | 7504 | 0955 | 33 |
| 28 | 5418 | 6543 | 4748 | 968 0018 | 2339 | 1699 | 8086 | 1487 | 32 |
| 29 | 6294 | 7371 | 5527 | 0748 | 3020 | 2330 | 8668 | 2019 | 31 |
| 30 | 7170 | 8197 | 6305 | 1476 | 3699 | 2960 | 9247 | 2549 | 30 |
| 31 | 8044 | 9023 | 7081 | 2204 | 4378 | 3589 | 9827 | 3079 | 29 |
| 32 | 8917 | 9848 | 7858 | 2931 | 5056 | 4218 | 980 0405 | 3608 | 28 |
| 33 | 9790 | 959 0672 | 8633 | 3658 | 5733 | 4845 | 0983 | 4136 | 27 |
| 34 | 954 0662 | 1496 | 9407 | 4383 | 6409 | 5472 | 1560 | 4663 | 26 |
| 35 | 1533 | 2318 | 964 0181 | 5108 | 7084 | 6098 | 2136 | 5189 | 25 |
| 36 | 2403 | 3140 | 0954 | 5832 | 7759 | 6723 | 2712 | 5715 | 24 |
| 37 | 3273 | 3961 | 1726 | 6555 | 8432 | 7347 | 3286 | 6239 | 23 |
| 38 | 4141 | 4781 | 2497 | 7277 | 9105 | 7970 | 3860 | 6763 | 22 |
| 39 | 5009 | 5600 | 3268 | 7998 | 9777 | 8593 | 4433 | 7286 | 21 |
| 40 | 5876 | 6418 | 4037 | 8719 | 973 0449 | 9215 | 5005 | 7808 | 20 |
| 41 | 6743 | 7236 | 4806 | 9438 | 1119 | 9836 | 5576 | 8330 | 19 |
| 42 | 7608 | 8053 | 5574 | 969 0157 | 1789 | 977 0456 | 6147 | 8850 | 18 |
| 43 | 8473 | 8869 | 6341 | 0875 | 2458 | 1075 | 6716 | 9370 | 17 |
| 44 | 9336 | 9684 | 7108 | 1593 | 3125 | 1693 | 7285 | 9889 | 16 |
| 45 | 955 0199 | 960 0499 | 7873 | 2309 | 3793 | 2311 | 7853 | 984 0407 | 15 |
| 46 | 1062 | 1312 | 8638 | 3025 | 4459 | 2928 | 8420 | 0924 | 14 |
| 47 | 1923 | 2125 | 9402 | 3740 | 5124 | 3544 | 8986 | 1441 | 13 |
| 48 | 2784 | 2937 | 965 0165 | 4453 | 5789 | 4159 | 9552 | 1956 | 12 |
| 49 | 3643 | 3748 | 0927 | 5167 | 6453 | 4773 | 981 0116 | 2471 | 11 |
| 50 | 4502 | 4558 | 1689 | 5879 | 7116 | 5387 | 0680 | 2985 | 10 |
| 51 | 5361 | 5368 | 2449 | 6591 | 7778 | 5999 | 1243 | 3498 | 9 |
| 52 | 6218 | 6177 | 3209 | 7301 | 8439 | 6611 | 1805 | 4010 | 8 |
| 53 | 7074 | 6984 | 3968 | 8011 | 9100 | 7222 | 2366 | 4521 | 7 |
| 54 | 7930 | 7792 | 4726 | 8720 | 9760 | 7832 | 2927 | 5032 | 6 |
| 55 | 8785 | 8598 | 5484 | 9428 | 974 0419 | 8442 | 3486 | 5542 | 5 |
| 56 | 9639 | 9403 | 6240 | 970 0136 | 1077 | 9050 | 4045 | 6050 | 4 |
| 57 | 956 0492 | 961 0208 | 6996 | 0842 | 1734 | 9658 | 4603 | 6558 | 3 |
| 58 | 1345 | 1012 | 7751 | 1548 | 2390 | 978 0265 | 5160 | 7066 | 2 |
| 59 | 2197 | 1815 | 8505 | 2253 | 3046 | 0871 | 5716 | 7572 | 1 |
| 60 | 3048 | 2617 | 9258 | 2957 | 3701 | 1476 | 6272 | 8078 | 0 |
| ′ | 17° | 16° | 15° | 14° | 13° | 12° | 11° | 10° | ′ |

NAT. COSINE.

| ′ | 72° | 73° | 74° | 75° | 76° | 77° | 78° | ′ |
|---|---|---|---|---|---|---|---|---|
| 0 | 3·0776835 | 3·2708526 | 3·4874144 | 3·7320508 | 4·0107809 | 4·3314759 | 4·7046301 | 60 |
| 1 | 3·0807325 | 42588 | 3·4912470 | 63980 | 57570 | 72316 | 4·7113686 | 59 |
| 2 | 37869 | 76715 | 50874 | 3·7407546 | 4·0207446 | 4·3430018 | 81256 | 58 |
| 3 | 68468 | 3·2810907 | 89356 | 51207 | 57440 | 87866 | 4·7249012 | 57 |
| 4 | 99122 | 45164 | 3·5027916 | 94963 | 4·0307550 | 4·3545861 | 4·7316954 | 56 |
| 5 | 3·0929831 | 79487 | 66555 | 3·7538815 | 57779 | 4·3604003 | 85083 | 55 |
| 6 | 60596 | 3·2913876 | 3·5105273 | 82763 | 4·0408125 | 62293 | 4·7453401 | 54 |
| 7 | 91416 | 48330 | 44070 | 3·7626807 | 58590 | 4·3720731 | 4·7521907 | 53 |
| 8 | 3·1022291 | 82851 | 82946 | 70947 | 4·0509174 | 79317 | 90603 | 52 |
| 9 | 53223 | 3·3017438 | 3·5221902 | 3·7715185 | 59877 | 4·3838054 | 4·7659490 | 51 |
| 10 | 84210 | 52091 | 60938 | 59519 | 4·0610700 | 96940 | 4·7728568 | 50 |
| 11 | 3·1115254 | 86811 | 3·5300054 | 3·7803951 | 61643 | 4·3955977 | 97837 | 49 |
| 12 | 46353 | 3·3121598 | 39251 | 48481 | 4·0712707 | 4·4015164 | 4·7867300 | 48 |
| 13 | 77509 | 56452 | 78528 | 93109 | 63892 | 74504 | 4·7936957 | 47 |
| 14 | 3·1208722 | 91373 | 3·5417886 | 3·7937835 | 4·0815199 | 4·4133996 | 4·8006808 | 46 |
| 15 | 39991 | 3·3226362 | 57325 | 82661 | 66627 | 93641 | 76854 | 45 |
| 16 | 71317 | 61419 | 96846 | 3·8027585 | 4·0918178 | 4·4253439 | 4·8147096 | 44 |
| 17 | 3·1302701 | 96543 | 3·5526449 | 72609 | 69852 | 4·4313392 | 4·8217536 | 43 |
| 18 | 34141 | 3·3331736 | 76133 | 3·8117733 | 4·1021649 | 73500 | 88174 | 42 |
| 19 | 65639 | 66997 | 3·5615900 | 62957 | 73569 | 4·4433762 | 4·8359010 | 41 |
| 20 | 97194 | 3·3402326 | 55749 | 3·8208281 | 4·1125614 | 94181 | 4·8430045 | 40 |
| 21 | 3·1428807 | 37724 | 95681 | 53707 | 77784 | 4·4554756 | 4·8501282 | 39 |
| 22 | 60478 | 73191 | 3·5735696 | 99233 | 4·1230079 | 4·4615489 | 72719 | 38 |
| 23 | 92207 | 3·3508728 | 75794 | 3·8344861 | 82499 | 76379 | 4·8644359 | 37 |
| 24 | 3·1523994 | 44333 | 3·5815975 | 90591 | 4·1335046 | 4·4737428 | 4·8716201 | 36 |
| 25 | 55840 | 80008 | 56241 | 3·8436424 | 87719 | 98636 | 88248 | 35 |
| 26 | 87744 | 3·3615753 | 96590 | 82358 | 4·1440519 | 4·4860004 | 4·8860499 | 34 |
| 27 | 3·1619706 | 51568 | 3·5937024 | 3·8528396 | 93446 | 4·4921532 | 4·8932956 | 33 |
| 28 | 51728 | 87453 | 77543 | 74537 | 4·1546501 | 83221 | 4·9005620 | 32 |
| 29 | 83808 | 3·3723408 | 3·6018146 | 3·8620782 | 99685 | 4·5045072 | 78491 | 31 |
| 30 | 3·1715948 | 59434 | 58835 | 67131 | 4·1652998 | 4·5107085 | 4·9151570 | 30 |
| 31 | 48147 | 95531 | 99609 | 3·8713584 | 4·1706440 | 69261 | 4·9224859 | 29 |
| 32 | 80406 | 3·3831699 | 3·6140469 | 60142 | 60011 | 4·5231601 | 98358 | 28 |
| 33 | 3·1812724 | 67938 | 81415 | 3·8806805 | 4·1813713 | 94105 | 4·9372068 | 27 |
| 34 | 45102 | 3·3904249 | 3·6222447 | 53574 | 67546 | 4·5356773 | 4·9445990 | 26 |
| 35 | 77540 | 40631 | 63566 | 3·8900448 | 4·1921510 | 4·5419608 | 4·9520125 | 25 |
| 36 | 3·1910039 | 77085 | 3·6304771 | 47429 | 75606 | 82608 | 94474 | 24 |
| 37 | 42598 | 3·4013612 | 46064 | 94516 | 4·2029835 | 4·5545776 | 4·9669037 | 23 |
| 38 | 75217 | 50210 | 87444 | 3·9041710 | 84196 | 4·5609111 | 4·9743817 | 22 |
| 39 | 3·2007897 | 86882 | 3·6428911 | 89011 | 4·2138690 | 72615 | 4·9818813 | 21 |
| 40 | 40638 | 3·4123626 | 70467 | 3·9136420 | 93318 | 4·5736287 | 94027 | 20 |
| 41 | 73440 | 60443 | 3·6512111 | 83937 | 4·2248080 | 4·5800129 | 4·9969459 | 19 |
| 42 | 3·2106304 | 97333 | 53844 | 3·9231563 | 4·2302977 | 64141 | 5·0045111 | 18 |
| 43 | 39228 | 3·4234297 | 95665 | 79297 | 58009 | 4·5928325 | 5·0120984 | 17 |
| 44 | 72215 | 71334 | 3·6637575 | 3·9327141 | 4·2413177 | 92680 | 97078 | 16 |
| 45 | 3·2205263 | 3·4308446 | 79575 | 75094 | 68482 | 4·6057207 | 5·0273395 | 15 |
| 46 | 38373 | 45631 | 3·6721665 | 3·9423157 | 4·2523923 | 4·6121908 | 5·0349935 | 14 |
| 47 | 71546 | 82891 | 63845 | 71331 | 79501 | 86783 | 5·0426700 | 13 |
| 48 | 3·2304780 | 3·4420226 | 3·6806115 | 3·9519615 | 4·2635218 | 4·6251832 | 5·0503690 | 12 |
| 49 | 38078 | 57635 | 48475 | 68011 | 91072 | 4·6317056 | 80907 | 11 |
| 50 | 71438 | 95120 | 90927 | 3·9616518 | 4·2747066 | 82457 | 5·0658352 | 10 |
| 51 | 3·2404860 | 3·4532679 | 3·6933469 | 65137 | 4·2803199 | 4·6448034 | 5·0736025 | 9 |
| 52 | 38346 | 70315 | 76104 | 3·9713868 | 59472 | 4·6513788 | 5·0813928 | 8 |
| 53 | 71895 | 3·4608026 | 3·7018830 | 62712 | 4·2915885 | 79721 | 92061 | 7 |
| 54 | 3·2505508 | 45813 | 61648 | 3·9811669 | 72440 | 4·6645832 | 5·0970426 | 6 |
| 55 | 39184 | 83676 | 3·7104558 | 60739 | 4·3029136 | 4·6712124 | 5·1049024 | 5 |
| 56 | 72924 | 3·4721616 | 47561 | 3·9909924 | 85974 | 78595 | 5·1127855 | 4 |
| 57 | 3·2606728 | 59632 | 90658 | 59223 | 4·3142955 | 4·6845248 | 5·1206921 | 3 |
| 58 | 40596 | 97726 | 3·7233847 | 4·0008636 | 4·3200079 | 4·6912083 | 86224 | 2 |
| 59 | 74529 | 3·4835896 | 77131 | 58165 | 57347 | 79100 | 5·1365763 | 1 |
| 60 | 3·2708526 | 74144 | 3·7320508 | 4·0107809 | 4·3314759 | 4·7046301 | 5·1445540 | 0 |
| ′ | 17° | 16° | 15° | 14° | 13° | 12° | 11° | ′ |

NAT. COTAN.

NAT. SINE.

| ′ | 80° | 81° | 82° | 83° | 84° | 85° | 86° | 87° | ′ |
|---|---|---|---|---|---|---|---|---|---|
| 0 | 9848 078 | 9876 883 | 9902 681 | 9925 462 | 9945 219 | 9961 947 | 9975 641 | 9986 295 | 60 |
| 1 | 582 | 9877 338 | 9903 085 | 816 | 523 | 9962 200 | 843 | 447 | 59 |
| 2 | 9849 086 | 792 | 489 | 9926 169 | 825 | 452 | 9976 045 | 598 | 58 |
| 3 | 589 | 9878 245 | 891 | 521 | 9946 127 | 704 | 245 | 748 | 57 |
| 4 | 9850 091 | 697 | 9904 293 | 873 | 428 | 954 | 445 | 898 | 56 |
| 5 | 593 | 9879 148 | 694 | 9927 224 | 729 | 9963 204 | 645 | 9987 046 | 55 |
| 6 | 9351 093 | 599 | 9905 095 | 573 | 9947 028 | 453 | 843 | 194 | 54 |
| 7 | 593 | 9880 048 | 494 | 922 | 327 | 701 | 9977 040 | 340 | 53 |
| 8 | 9852 092 | 497 | 893 | 9928 271 | 625 | 948 | 237 | 486 | 52 |
| 9 | 590 | 945 | 9906 290 | 618 | 921 | 9964 195 | 433 | 631 | 51 |
| 10 | 9853 087 | 9881 392 | 687 | 965 | 9948 217 | 440 | 627 | 775 | 50 |
| 11 | 583 | 838 | 9907 083 | 9929 310 | 513 | 685 | 821 | 919 | 49 |
| 12 | 9854 079 | 9882 284 | 478 | 655 | 807 | 929 | 9978 015 | 9988 061 | 48 |
| 13 | 574 | 728 | 873 | 999 | 9949 101 | 9965 172 | 207 | 203 | 47 |
| 14 | 9855 068 | 9883 172 | 9908 266 | 9930 342 | 393 | 414 | 399 | 344 | 46 |
| 15 | 561 | 615 | 659 | 685 | 685 | 655 | 589 | 484 | 45 |
| 16 | 9856 053 | 9884 057 | 9909 051 | 9931 026 | 976 | 895 | 779 | 623 | 44 |
| 17 | 544 | 498 | 442 | 367 | 9950 266 | 9966 135 | 968 | 761 | 43 |
| 18 | 9857 035 | 939 | 832 | 706 | 556 | 374 | 9979 156 | 899 | 42 |
| 19 | 524 | 9885 378 | 9910 221 | 9932 045 | 844 | 612 | 343 | 9989 035 | 41 |
| 20 | 9858 013 | 817 | 610 | 384 | 9951 132 | 849 | 530 | 171 | 40 |
| 21 | 501 | 9886 255 | 997 | 721 | 419 | 9967 085 | 716 | 306 | 39 |
| 22 | 988 | 692 | 9911 384 | 9933 057 | 705 | 321 | 900 | 440 | 38 |
| 23 | 9859 475 | 9887 128 | 770 | 393 | 990 | 555 | 9980 084 | 573 | 37 |
| 24 | 960 | 564 | 9912 155 | 728 | 9952 274 | 789 | 267 | 706 | 36 |
| 25 | 9860 445 | 998 | 540 | 9934 062 | 557 | 9968 022 | 450 | 837 | 35 |
| 26 | 929 | 9888 432 | 923 | 395 | 840 | 254 | 631 | 968 | 34 |
| 27 | 9861 412 | 865 | 9913 306 | 727 | 9953 122 | 485 | 811 | 9990 098 | 33 |
| 28 | 894 | 9889 297 | 688 | 9935 058 | 403 | 715 | 991 | 227 | 32 |
| 29 | 9862 375 | 728 | 9914 069 | 389 | 683 | 945 | 9981 170 | 355 | 31 |
| 30 | 856 | 9890 159 | 449 | 719 | 962 | 9969 173 | 348 | 482 | 30 |
| 31 | 9863 336 | 588 | 828 | 9936 047 | 9954 240 | 401 | 525 | 609 | 29 |
| 32 | 815 | 9891 017 | 9915 206 | 375 | 518 | 628 | 701 | 734 | 28 |
| 33 | 9864 293 | 445 | 584 | 703 | 795 | 854 | 877 | 859 | 27 |
| 34 | 770 | 872 | 961 | 9937 029 | 9955 070 | 9970 080 | 9982 052 | 983 | 26 |
| 35 | 9865 246 | 9892 298 | 9916 337 | 355 | 345 | 304 | 225 | 9991 106 | 25 |
| 36 | 722 | 723 | 712 | 679 | 620 | 528 | 398 | 228 | 24 |
| 37 | 9866 196 | 9893 148 | 9917 086 | 9938 003 | 893 | 750 | 570 | 350 | 23 |
| 38 | 670 | 572 | 459 | 326 | 9956 165 | 972 | 742 | 470 | 22 |
| 39 | 9867 143 | 994 | 832 | 648 | 437 | 9971 193 | 912 | 590 | 21 |
| 40 | 615 | 9894 416 | 9918 204 | 969 | 708 | 413 | 9983 082 | 709 | 20 |
| 41 | 9868 087 | 838 | 574 | 9939 290 | 978 | 633 | 250 | 827 | 19 |
| 42 | 557 | 9895 258 | 944 | 610 | 9957 247 | 851 | 418 | 944 | 18 |
| 43 | 9869 027 | 677 | 9919 314 | 928 | 515 | 9972 069 | 585 | 9992 060 | 17 |
| 44 | 496 | 9896 096 | 682 | 9940 246 | 783 | 286 | 751 | 176 | 16 |
| 45 | 964 | 514 | 9920 049 | 563 | 9958 049 | 502 | 917 | 290 | 15 |
| 46 | 9870 431 | 931 | 416 | 880 | 315 | 717 | 9984 081 | 404 | 14 |
| 47 | 897 | 9897 347 | 782 | 9941 195 | 580 | 931 | 245 | 517 | 13 |
| 48 | 9871 363 | 762 | 9921 147 | 510 | 844 | 9973 145 | 408 | 629 | 12 |
| 49 | 827 | 9898 177 | 511 | 823 | 9959 107 | 357 | 570 | 740 | 11 |
| 50 | 9872 291 | 590 | 874 | 9942 136 | 370 | 569 | 731 | 851 | 10 |
| 51 | 764 | 9899 003 | 9922 237 | 448 | 631 | 780 | 891 | 960 | 9 |
| 52 | 9873 216 | 415 | 599 | 760 | 892 | 990 | 9985 050 | 9993 069 | 8 |
| 53 | 678 | 826 | 959 | 9943 070 | 9960 152 | 9974 199 | 209 | 177 | 7 |
| 54 | 9874 138 | 9900 237 | 9923 319 | 379 | 411 | 408 | 367 | 284 | 6 |
| 55 | 598 | 646 | 679 | 688 | 669 | 615 | 524 | 390 | 5 |
| 56 | 9875 057 | 9901 055 | 9924 037 | 996 | 926 | 822 | 680 | 195 | 4 |
| 57 | 514 | 462 | 394 | 9944 303 | 9961 183 | 9975 028 | 835 | 600 | 3 |
| 58 | 972 | 869 | 751 | 609 | 438 | 233 | 989 | 704 | 2 |
| 59 | 9876 428 | 9902 275 | 9925 107 | 914 | 693 | 437 | 9986 143 | 806 | 1 |
| 60 | 883 | 681 | 462 | 9945 219 | 947 | 641 | 295 | 908 | 0 |
| ′ | 9° | 8° | 7° | 6° | 5° | 4° | 3° | 2° | ′ |

NAT. COSINE.

| ′ | 79° | 80° | 81° | 82° | 83° | 84° | 85° | ′ |
|---|---|---|---|---|---|---|---|---|
| 0 | 5·1445540 | 5·6712818 | 6·3137515 | 7·1153697 | 8·1443464 | 9·5143645 | 11·430052 | 60 |
| 1 | 525557 | 809446 | 256601 | 304190 | 639786 | 410613 | 468474 | 59 |
| 2 | 605813 | 906394 | 376126 | 455308 | 837041 | 679068 | 507154 | 58 |
| 3 | 686311 | 5·7003663 | 496092 | 607056 | 8·2035239 | 949022 | 546093 | 57 |
| 4 | 767051 | 101256 | 616502 | 759437 | 234384 | 9·6220486 | 585294 | 56 |
| 5 | 848035 | 199173 | 737359 | 912456 | 434485 | 493475 | 624761 | 55 |
| 6 | 929264 | 297416 | 858665 | 7·2066116 | 635547 | 768000 | 664495 | 54 |
| 7 | 5·2010738 | 395988 | 980422 | 220422 | 837579 | 9·7044075 | 704500 | 53 |
| 8 | 092459 | 494889 | 6·4102633 | 375378 | 8·3040586 | 321713 | 744779 | 52 |
| 9 | 174428 | 594122 | 225301 | 530987 | 244577 | 600927 | 785333 | 51 |
| 10 | 256647 | 693688 | 348428 | 687255 | 449558 | 881732 | 826167 | 50 |
| 11 | 339116 | 793588 | 472017 | 844184 | 655536 | 9·8164140 | 867282 | 49 |
| 12 | 421836 | 893825 | 596070 | 7·3001780 | 862519 | 448166 | 908682 | 48 |
| 13 | 504809 | 994400 | 720591 | 160047 | 8·4070515 | 733823 | 950370 | 47 |
| 14 | 588035 | 5·8095315 | 845581 | 318989 | 279531 | 9·9021125 | 992349 | 46 |
| 15 | 671517 | 196572 | 971043 | 478610 | 489573 | 310088 | 12·034622 | 45 |
| 16 | 755255 | 298172 | 6·5096981 | 638916 | 700651 | 600724 | 077192 | 44 |
| 17 | 839251 | 400117 | 223396 | 799909 | 912772 | 893050 | 120062 | 43 |
| 18 | 923505 | 502410 | 350293 | 961595 | 8·5125943 | 10·018708 | 163236 | 42 |
| 19 | 5·3008018 | 605051 | 477672 | 7·4123978 | 340172 | 048283 | 206716 | 41 |
| 20 | 092793 | 708042 | 605538 | 287064 | 555468 | 078031 | 250505 | 40 |
| 21 | 177830 | 811386 | 733892 | 450855 | 771838 | 107954 | 294609 | 39 |
| 22 | 263131 | 915084 | 862739 | 615357 | 989290 | 138054 | 339028 | 38 |
| 23 | 348696 | 5·9019138 | 992080 | 780576 | 8·6207833 | 168332 | 383768 | 37 |
| 24 | 434527 | 123550 | 6·6121919 | 946514 | 427475 | 198789 | 428831 | 36 |
| 25 | 520626 | 228322 | 252258 | 7·5113178 | 648223 | 229428 | 474221 | 35 |
| 26 | 606993 | 333455 | 383100 | 280571 | 870088 | 260249 | 519942 | 34 |
| 27 | 693630 | 438952 | 514449 | 448699 | 8·7093077 | 291255 | 565997 | 33 |
| 28 | 780538 | 544815 | 646307 | 617567 | 317198 | 322447 | 612390 | 32 |
| 29 | 867718 | 651045 | 778677 | 787179 | 542461 | 353827 | 659125 | 31 |
| 30 | 955172 | 757644 | 911562 | 957541 | 768874 | 385397 | 706205 | 30 |
| 31 | 5·4042901 | 864614 | 6·7044966 | 7·6128657 | 996446 | 417158 | 753634 | 29 |
| 32 | 130906 | 971957 | 178891 | 300533 | 8·8225186 | 449112 | 801417 | 28 |
| 33 | 219188 | 6·0079676 | 313341 | 473174 | 455103 | 481261 | 849557 | 27 |
| 34 | 307750 | 187772 | 448318 | 646584 | 686206 | 513607 | 898058 | 26 |
| 35 | 396592 | 296247 | 583826 | 820769 | 918505 | 546151 | 946924 | 25 |
| 36 | 485715 | 405103 | 719867 | 995735 | 8·9152009 | 578895 | 996160 | 24 |
| 37 | 575121 | 514343 | 856446 | 7·7171486 | 386726 | 611841 | 13·045769 | 23 |
| 38 | 664812 | 623967 | 993565 | 348028 | 622668 | 644992 | 095757 | 22 |
| 39 | 754788 | 733979 | 6·8131227 | 525366 | 859843 | 678348 | 146127 | 21 |
| 40 | 845052 | 844381 | 269437 | 703506 | 9·0098261 | 711913 | 196883 | 20 |
| 41 | 935604 | 955174 | 408196 | 882453 | 337933 | 745687 | 248031 | 19 |
| 42 | 5·5026446 | 6·1066360 | 547508 | 7·8062212 | 578867 | 779673 | 299574 | 18 |
| 43 | 117579 | 177943 | 687378 | 242790 | 821074 | 813872 | 351518 | 17 |
| 44 | 209005 | 289923 | 827807 | 424191 | 9·1064564 | 848288 | 403867 | 16 |
| 45 | 300724 | 402303 | 968799 | 606423 | 309348 | 882921 | 456625 | 15 |
| 46 | 392740 | 515085 | 6·9110359 | 789489 | 555436 | 917775 | 599799 | 14 |
| 47 | 485052 | 628272 | 252489 | 973396 | 802838 | 952850 | 563391 | 13 |
| 48 | 577663 | 741865 | 395192 | 7·9158151 | 9·2051564 | 988150 | 617409 | 12 |
| 49 | 670574 | 855867 | 538473 | 343758 | 301627 | 11·023676 | 671856 | 11 |
| 50 | 763786 | 970279 | 682335 | 530224 | 553035 | 059431 | 726738 | 10 |
| 51 | 857302 | 6·2085106 | 826781 | 717555 | 805802 | 095416 | 782060 | 9 |
| 52 | 951121 | 200347 | 971806 | 905756 | 9·3059936 | 131635 | 837827 | 8 |
| 53 | 5·6045247 | 316007 | 7·0117441 | 8·0094835 | 315450 | 168089 | 894045 | 7 |
| 54 | 139680 | 432086 | 263662 | 284796 | 572355 | 204780 | 950719 | 6 |
| 55 | 234421 | 548588 | 410482 | 475647 | 830663 | 241712 | 14·007856 | 5 |
| 56 | 329474 | 665515 | 557905 | 667394 | 9·4090384 | 278885 | 065459 | 4 |
| 57 | 424838 | 782868 | 705934 | 860042 | 351531 | 316304 | 123536 | 3 |
| 58 | 520516 | 900651 | 854573 | 8·1053599 | 614116 | 353970 | 182092 | 2 |
| 59 | 616509 | 6·3018866 | 7·1003826 | 248071 | 878149 | 391885 | 241134 | 1 |
| 60 | 712818 | 137515 | 153697 | 443464 | 9·5143645 | 430052 | 300666 | 0 |
| ′ | 10° | 9° | 8° | 7° | 6° | 5° | 4° | ′ |

NAT. COTAN.

NAT. SINE.

| ′ | 88° | 89° | ′ |
|---|---|---|---|
| 0 | 9993 908 | 9998 477 | 60 |
| 1 | 9994 009 | 527 | 59 |
| 2 | 110 | 577 | 58 |
| 3 | 209 | 625 | 57 |
| 4 | 308 | 673 | 56 |
| 5 | 405 | 720 | 55 |
| 6 | 502 | 766 | 54 |
| 7 | 598 | 812 | 53 |
| 8 | 693 | 856 | 52 |
| 9 | 788 | 900 | 51 |
| 10 | 881 | 942 | 50 |
| 11 | 974 | 984 | 49 |
| 12 | 9995 066 | 9999 025 | 48 |
| 13 | 157 | 065 | 47 |
| 14 | 247 | 105 | 46 |
| 15 | 336 | 143 | 45 |
| 16 | 424 | 181 | 44 |
| 17 | 512 | 218 | 43 |
| 18 | 599 | 254 | 42 |
| 19 | 684 | 289 | 41 |
| 20 | 770 | 323 | 40 |
| 21 | 854 | 357 | 39 |
| 22 | 937 | 389 | 38 |
| 23 | 9996 020 | 421 | 37 |
| 24 | 101 | 452 | 36 |
| 25 | 182 | 482 | 35 |
| 26 | 262 | 511 | 34 |
| 27 | 341 | 539 | 33 |
| 28 | 419 | 567 | 32 |
| 29 | 497 | 593 | 31 |
| 30 | 573 | 619 | 30 |
| 31 | 649 | 644 | 29 |
| 32 | 724 | 668 | 28 |
| 33 | 798 | 602 | 27 |
| 34 | 871 | 714 | 26 |
| 35 | 943 | 736 | 25 |
| 36 | 9997 015 | 756 | 24 |
| 37 | 086 | 776 | 23 |
| 38 | 156 | 795 | 22 |
| 39 | 224 | 813 | 21 |
| 40 | 292 | 831 | 20 |
| 41 | 360 | 847 | 19 |
| 42 | 426 | 863 | 18 |
| 43 | 492 | 878 | 17 |
| 44 | 556 | 892 | 16 |
| 45 | 620 | 905 | 15 |
| 46 | 683 | 917 | 14 |
| 47 | 745 | 928 | 13 |
| 48 | 807 | 939 | 12 |
| 49 | 867 | 949 | 11 |
| 50 | 927 | 958 | 10 |
| 51 | 986 | 966 | 9 |
| 52 | 9998 044 | 973 | 8 |
| 53 | 101 | 979 | 7 |
| 54 | 157 | 985 | 6 |
| 55 | 213 | 989 | 5 |
| 56 | 267 | 993 | 4 |
| 57 | 321 | 996 | 3 |
| 58 | 374 | 998 | 2 |
| 59 | 426 | 1·0000 000 | 1 |
| 60 | 477 | 000 | 0 |
| ′ | 1° | 0° | ′ |

NAT. COSINE.

NAT. TAN.

| ′ | 86° | 87° | 88° | 89° | ′ |
|---|---|---|---|---|---|
| 0 | 14·300666 | 19·081137 | 28·636253 | 57·289962 | 60 |
| 1 | 360696 | 187930 | 877089 | 58·261174 | 59 |
| 2 | 421230 | 295922 | 29·122005 | 59·265872 | 58 |
| 3 | 482273 | 405133 | 371106 | 60·305820 | 57 |
| 4 | 543833 | 515584 | 624499 | 61·382905 | 56 |
| 5 | 605916 | 627296 | 882299 | 62·499154 | 55 |
| 6 | 668529 | 740291 | 30·144619 | 63·656741 | 54 |
| 7 | 731679 | 854591 | 411580 | 64·858008 | 53 |
| 8 | 795372 | 970219 | 683307 | 66·105473 | 52 |
| 9 | 859616 | 20·087199 | 959928 | 67·401854 | 51 |
| 10 | 924417 | 205553 | 31·241577 | 68·750087 | 50 |
| 11 | 989784 | 325308 | 528392 | 70·153346 | 49 |
| 12 | 15·055723 | 446486 | 820516 | 71·615070 | 48 |
| 13 | 122242 | 569115 | 32·118099 | 73·138991 | 47 |
| 14 | 189349 | 693220 | 421295 | 74·729165 | 46 |
| 15 | 257052 | 818828 | 730264 | 76·390009 | 45 |
| 16 | 325358 | 945966 | 33·045173 | 78·126342 | 44 |
| 17 | 394276 | 21·074664 | 366194 | 79·943430 | 43 |
| 18 | 463814 | 204949 | 693509 | 81·847041 | 42 |
| 19 | 533981 | 336851 | 34·027303 | 83·843507 | 41 |
| 20 | 604784 | 470401 | 367771 | 85·939791 | 40 |
| 21 | 676233 | 605630 | 715115 | 88·143572 | 39 |
| 22 | 748337 | 742569 | 35·069546 | 90·463336 | 38 |
| 23 | 821105 | 881251 | 431282 | 92·908487 | 37 |
| 24 | 894545 | 22·021710 | 800553 | 95·489475 | 36 |
| 25 | 968667 | 163980 | 36·177596 | 98·217943 | 35 |
| 26 | 16·043482 | 308097 | 562659 | 101·10690 | 34 |
| 27 | 118998 | 454096 | 956001 | 104·17094 | 33 |
| 28 | 195225 | 602015 | 37·357892 | 107·42648 | 32 |
| 29 | 272174 | 751892 | 768613 | 110·89205 | 31 |
| 30 | 349855 | 903766 | 38·188459 | 114·58865 | 30 |
| 31 | 428279 | 23·057677 | 617738 | 118·54018 | 29 |
| 32 | 507456 | 213666 | 39·056771 | 122·77396 | 28 |
| 33 | 587396 | 371777 | 505895 | 127·32134 | 27 |
| 34 | 668112 | 532052 | 965460 | 132·21851 | 26 |
| 35 | 749614 | 694537 | 40·435837 | 137·50745 | 25 |
| 36 | 831915 | 859277 | 917412 | 143·23712 | 24 |
| 37 | 915025 | 24·026320 | 41·410588 | 149·46502 | 23 |
| 38 | 998957 | 195714 | 915790 | 156·25908 | 22 |
| 39 | 17·083724 | 367509 | 42·433464 | 163·70019 | 21 |
| 40 | 169337 | 541758 | 964077 | 171·88540 | 20 |
| 41 | 255809 | 718512 | 43·508122 | 180·93220 | 19 |
| 42 | 343155 | 897826 | 44·066113 | 190·98419 | 18 |
| 43 | 431385 | 25·079757 | 638596 | 202·21875 | 17 |
| 44 | 520516 | 264361 | 45·226141 | 214·85762 | 16 |
| 45 | 610559 | 451700 | 829351 | 229·18166 | 15 |
| 46 | 701529 | 641832 | 46·448862 | 245·55198 | 14 |
| 47 | 793442 | 834823 | 47·085343 | 264·44080 | 13 |
| 48 | 886310 | 26·030736 | 739501 | 286·47773 | 12 |
| 49 | 980150 | 229638 | 48·412084 | 312·52137 | 11 |
| 50 | 18·074977 | 431600 | 49·103881 | 343·77371 | 10 |
| 51 | 170807 | 636690 | 815726 | 381·97099 | 9 |
| 52 | 267654 | 844984 | 50·548506 | 429·71757 | 8 |
| 53 | 365537 | 27·056557 | 51·303157 | 491·10600 | 7 |
| 54 | 464471 | 271486 | 52·080673 | 572·95721 | 6 |
| 55 | 564473 | 489853 | 882109 | 687·54687 | 5 |
| 56 | 665562 | 711740 | 53·708587 | 859·43630 | 4 |
| 57 | 767754 | 937233 | 54·561300 | 1145·9153 | 3 |
| 58 | 871068 | 28·166422 | 55·441517 | 1718·8732 | 2 |
| 59 | 975523 | 399397 | 56·350590 | 3437·7467 | 1 |
| 60 | 19·081137 | 636253 | 57·289962 | Infinite. | 0 |
| ′ | 3° | 2° | 1° | 0° | ′ |

NAT. COTAN.

## TABLE OF RADII, &c.—Chord 100 Feet.

| Angle of Deflex'n. | | Radius in feet. | Def. dist. in feet. | Angle of Deflex'n. | | Radius in feet. | Def. dist. in feet. | Angle of Deflex'n. | | Radius in feet. | Def. dist. in feet. |
|---|---|---|---|---|---|---|---|---|---|---|---|
| ° | ′ | | | ° | ′ | | | ° | ′ | | |
| | 5 | 68760·0 | ·145 | 3 | 25 | 1677·0 | 5·962 | 12 | 15 | 468·7 | 21.360 |
| | 10 | 34380·0 | ·291 | | 30 | 1637·0 | 6·108 | | 30 | 459·3 | 21·790 |
| | 15 | 22920·0 | ·436 | | 35 | 1599·0 | 6·253 | | 45 | 450·3 | 22·210 |
| | 20 | 17190·0 | ·581 | | 40 | 1563·0 | 6·398 | 13 | | 441·7 | 22·640 |
| | 25 | 13752·0 | ·727 | | 45 | 1528·0 | 6·544 | | 15 | 433·4 | 23·070 |
| | 30 | 11460·0 | ·872 | | 50 | 1495·0 | 6·689 | | 30 | 425·5 | 23·510 |
| | 35 | 9823·0 | 1·017 | | 55 | 1463·0 | 6·835 | | 45 | 417·7 | 23·940 |
| | 40 | 8595·0 | 1·163 | 4 | | 1433.0 | 6·980 | 14 | | 410·3 | 24·370 |
| | 45 | 7640·0 | 1·308 | | 15 | 1348·0 | 7·416 | | 15 | 403·1 | 24·810 |
| | 50 | 6876·0 | 1·453 | | 30 | 1274·0 | 7·853 | | 30 | 396·2 | 25·240 |
| | 55 | 6251·0 | 1·600 | | 45 | 1207·0 | 8·289 | | 45 | 389·6 | 25·670 |
| 1 | | 5730·0 | 1·745 | 5 | | 1146·0 | 8·722 | 15 | | 383·1 | 26·110 |
| | 5 | 5289·0 | 1·890 | | 15 | 1092·0 | 9·159 | | 15 | 376·9 | 26·520 |
| | 10 | 4912·0 | 2·036 | | 30 | 1042·0 | 9·595 | | 30 | 370·8 | 26·940 |
| | 15 | 4584·0 | 2·181 | | 45 | 996·8 | 10·030 | | 45 | 365·0 | 27·370 |
| | 20 | 4298·0 | 2·327 | 6 | | 955·4 | 10·470 | 16 | | 359·3 | 27·830 |
| | 25 | 4045·0 | 2·472 | | 15 | 917·0 | 10·900 | | 30 | 348·4 | 28·700 |
| | 30 | 3820·0 | 2·618 | | 30 | 882·0 | 11·340 | 17 | | 338·3 | 29·560 |
| | 35 | 3619·0 | 2·763 | | 45 | 849·3 | 11·780 | | 30 | 328·7 | 30·430 |
| | 40 | 3438·0 | 2·908 | 7 | | 819·0 | 12·210 | 18 | | 319·6 | 31·290 |
| | 45 | 3274·0 | 3·054 | | 15 | 790·8 | 12·640 | | 30 | 311·0 | 32·500 |
| | 50 | 3125·0 | 3·199 | | 30 | 764·5 | 13·080 | 19 | | 302·9 | 33·010 |
| | 55 | 2990·0 | 3·345 | | 45 | 739·9 | 13·510 | | 30 | 295·3 | 33·870 |
| 2 | | 2865·0 | 3·490 | 8 | | 716·8 | 13·950 | 20 | | 287·9 | 34·730 |
| | 5 | 2750·0 | 3·635 | | 15 | 695·1 | 14·380 | 21 | | 274.4 | 36·440 |
| | 10 | 2644·0 | 3·781 | | 30 | 674·6 | 14·810 | 22 | | 262·0 | 38·150 |
| | 15 | 2547·0 | 3·926 | | 45 | 655·5 | 15·250 | 23 | | 250·8 | 39·870 |
| | 20 | 2456·0 | 4·072 | 9 | | 637·3 | 15·680 | 24 | | 240·5 | 41·580 |
| | 25 | 2371·0 | 4·217 | | 15 | 620·2 | 16·120 | 25 | | 231·0 | 43·280 |
| | 30 | 2292·0 | 4·363 | | 30 | 603·8 | 16·550 | 26 | | 222·3 | 44·980 |
| | 35 | 2218·0 | 4·508 | | 45 | 588·4 | 16·990 | 27 | | 214·2 | 46·680 |
| | 40 | 2149·0 | 4·653 | 10 | | 573·7 | 17·430 | 28 | | 206·7 | 48·380 |
| | 45 | 2084·0 | 4·799 | | 15 | 559·7 | 17·870 | 29 | | 199·7 | 50·070 |
| | 50 | 2023·0 | 4·944 | | 30 | 546.4 | 18·300 | 30 | | 193·2 | 51·760 |
| | 55 | 1965·0 | 5·090 | | 45 | 533·8 | 18·730 | 31 | | 187·1 | 53·450 |
| 3 | | 1910·0 | 5·235 | 11 | | 521·7 | 19·170 | 32 | | 181·4 | 55·130 |
| | 5 | 1859·0 | 5·380 | | 15 | 510·1 | 19·610 | 33 | | 176·0 | 56·800 |
| | 10 | 1810·0 | 5·526 | | 30 | 499·1 | 20·050 | 34 | | 171·0 | 58·470 |
| | 15 | 1763·0 | 5·671 | | 45 | 488·5 | 20·500 | 35 | | 166·3 | 60·140 |
| | 20 | 1719·0 | 5·817 | 12 | | 478·3 | 20·940 | 36 | | 161·8 | 61·800 |

# TABLE OF LONG CHORDS.

| Radius in feet. | Angle of Deflection. | Length of Chord in feet required to subtend | | | |
|---|---|---|---|---|---|
| | | 1 Station. | 2 Stations. | 3 Stations. | 4 Stations. |
| 5730·0 | 1° | 100 | 200·0 | 300·0 | 400·0 |
| 4584·0 | ¼ | 100 | 200·0 | 300·0 | 399·9 |
| 3820·0 | ½ | 100 | 200·0 | 300·0 | 399·9 |
| 3274·0 | ¾ | 100 | 200·0 | 300·0 | 399·8 |
| 2865·0 | 2° | 100 | 200·0 | 299·9 | 399·7 |
| 2547·0 | ¼ | 100 | 200·0 | 299·9 | 399·6 |
| 2292·0 | ½ | 100 | 200·0 | 299·8 | 399·5 |
| 2084·0 | ¾ | 100 | 200·0 | 299·8 | 399·4 |
| 1910·0 | 3° | 100 | 200·0 | 299·7 | 399·3 |
| 1763·0 | ¼ | 100 | 200·0 | 299·7 | 399·2 |
| 1637·0 | ½ | 100 | 200·0 | 299·6 | 399·1 |
| 1528·0 | ¾ | 100 | 200·0 | 299·6 | 399·0 |
| 1433·0 | 4° | 100 | 199·9 | 299·6 | 398·9 |
| 1348·0 | ¼ | 100 | 199·9 | 299·5 | 398·7 |
| 1274·0 | ½ | 100 | 199·9 | 299·4 | 398·5 |
| 1207·0 | ¾ | 100 | 199·9 | 299·3 | 398·3 |
| 1146·0 | 5° | 100 | 199·9 | 299·2 | 398·0 |
| 1092·0 | ¼ | 100 | 199·8 | 299·1 | 397·8 |
| 1042·0 | ½ | 100 | 199·8 | 299·0 | 397·6 |
| 996·8 | ¾ | 100 | 199·7 | 298·9 | 397·5 |
| 955.4 | 6° | 100 | 199·7 | 298·8 | 397·3 |
| 917·0 | ¼ | 100 | 199·7 | 298·7 | 397·0 |
| 882·0 | ½ | 100 | 199·7 | 298·6 | 396·7 |
| 849·3 | ¾ | 100 | 199·6 | 298·5 | 396·5 |
| 819·0 | 7° | 100 | 199·6 | 298·4 | 396·2 |
| 790·8 | ¼ | 100 | 199·6 | 298·3 | 396·0 |
| 764·5 | ½ | 100 | 199·6 | 298·2 | 395·7 |
| 739·9 | ¾ | 100 | 199·6 | 298·1 | 395·4 |
| 716·8 | 8° | 100 | 199·6 | 298·0 | 395·1 |
| 695·1 | ¼ | 100 | 199·5 | 297·9 | 394·8 |
| 674·6 | ½ | 100 | 199·5 | 297·8 | 394·5 |
| 655·5 | ¾ | 100 | 199·4 | 297·7 | 394·3 |
| 637·3 | 9° | 100 | 199·4 | 297·5 | 394·1 |
| 620·2 | ¼ | 100 | 199·4 | 297·4 | 393·7 |
| 603·8 | ½ | 100 | 199·3 | 297·3 | 393·2 |
| 588·4 | ¾ | 100 | 199·2 | 297·2 | 392·8 |
| 573·7 | 10° | 100 | 199·2 | 297·0 | 392·4 |

# TABLE OF ORDINATES.

## Ordinates 10 feet apart.—Chord 100 feet.

| Deflexion Angle in Degrees and Minutes. | Distances of the Ordinates from the end of the 100 feet Chord. | | | | |
|---|---|---|---|---|---|
| | 50 feet. | 40 feet. | 30 feet. | 20 feet. | 10 feet. |
| | Lengths of Ordinates in feet. | | | | |
| ° ′ | | | | | |
| 5 | ·018 | ·017 | ·015 | ·012 | ·006 |
| 10 | ·036 | ·035 | ·031 | ·023 | ·013 |
| 15 | ·054 | ·052 | ·046 | ·035 | ·019 |
| 20 | ·073 | ·070 | ·061 | ·047 | ·026 |
| 25 | ·091 | ·087 | ·076 | ·058 | ·032 |
| 30 | ·109 | ·105 | ·092 | ·070 | ·039 |
| 35 | ·127 | ·123 | ·108 | ·082 | ·045 |
| 40 | ·145 | ·140 | ·123 | ·093 | ·052 |
| 45 | ·163 | ·157 | ·137 | ·105 | ·058 |
| 50 | ·182 | ·175 | ·153 | ·117 | ·065 |
| 55 | ·200 | ·192 | ·168 | ·128 | ·071 |
| 1° | ·218 | ·209 | ·183 | ·140 | ·078 |
| 5 | ·236 | ·226 | ·198 | ·152 | ·085 |
| 10 | ·254 | ·244 | ·214 | ·163 | ·091 |
| 15 | ·273 | ·261 | ·229 | ·175 | ·098 |
| 20 | ·291 | ·279 | ·244 | ·187 | ·104 |
| 25 | ·309 | ·296 | ·259 | ·198 | ·111 |
| 30 | ·327 | ·314 | ·275 | ·210 | ·117 |
| 35 | ·345 | ·331 | ·290 | ·221 | ·124 |
| 40 | ·364 | ·349 | ·305 | ·233 | ·130 |
| 45 | ·382 | ·366 | ·321 | ·245 | ·137 |
| 50 | ·400 | ·384 | ·336 | ·256 | ·144 |
| 55 | ·418 | ·401 | ·351 | ·268 | ·150 |
| 2° | ·436 | ·419 | ·366 | ·280 | ·157 |
| 5 | ·454 | ·436 | ·382 | ·291 | ·163 |
| 10 | ·473 | ·454 | ·397 | ·303 | ·170 |
| 15 | ·491 | ·471 | ·412 | ·315 | ·176 |
| 20 | ·509 | ·489 | ·428 | ·326 | ·183 |
| 25 | ·527 | ·506 | ·443 | ·338 | ·190 |
| 30 | ·545 | ·524 | ·458 | ·350 | ·196 |
| 35 | ·564 | ·541 | ·474 | ·361 | ·203 |
| 40 | ·582 | ·559 | ·489 | ·373 | ·209 |
| 45 | ·600 | ·576 | ·504 | ·384 | ·216 |
| 50 | ·618 | ·594 | ·519 | ·396 | ·222 |
| 55 | ·636 | ·611 | ·535 | ·408 | ·229 |
| 3° | ·654 | ·629 | ·550 | ·419 | ·235 |
| 5 | ·673 | ·646 | ·565 | ·431 | ·242 |
| 10 | ·691 | ·664 | ·581 | ·443 | ·249 |
| 15 | ·709 | ·681 | ·596 | ·454 | ·255 |
| 20 | ·727 | ·699 | ·611 | ·466 | ·262 |
| 25 | ·745 | ·716 | ·627 | ·478 | ·268 |

## TABLE OF ORDINATES.—Continued.

Ordinates 10 feet apart.—Chord 100 feet.

| Deflexion Angle in Degrees and Minutes. | | Distances of the Ordinates from the end of the 100 feet Chord. | | | | |
|---|---|---|---|---|---|---|
| | | 50 feet. | 40 feet. | 30 feet. | 20 feet. | 10 feet. |
| | | Lengths of Ordinates in feet. | | | | |
| ° | ′ | | | | | |
| 3 | 30 | ·764 | ·734 | ·642 | ·489 | ·275 |
| | 35 | ·782 | ·751 | ·657 | ·501 | ·281 |
| | 40 | ·800 | ·769 | ·673 | ·512 | ·288 |
| | 45 | ·818 | ·786 | ·688 | ·524 | ·294 |
| | 50 | ·836 | ·804 | ·703 | ·536 | ·301 |
| | 55 | ·854 | ·821 | ·718 | ·547 | ·308 |
| 4 | | ·873 | ·839 | ·734 | ·559 | ·314 |
| | 15 | ·927 | ·891 | ·780 | ·594 | ·334 |
| | 30 | ·981 | ·944 | ·825 | ·629 | ·354 |
| | 45 | 1·036 | ·996 | ·871 | ·664 | ·373 |
| 5 | | 1·091 | 1·048 | ·917 | ·699 | ·393 |
| | 15 | 1·146 | 1·100 | ·963 | ·734 | ·413 |
| | 30 | 1·200 | 1·153 | 1·009 | ·769 | ·432 |
| | 45 | 1·255 | 1·205 | 1·055 | ·804 | ·452 |
| 6 | | 1·309 | 1·258 | 1·100 | ·839 | ·472 |
| | 15 | 1·364 | 1·310 | 1·146 | ·874 | ·492 |
| | 30 | 1·419 | 1·362 | 1·192 | ·909 | ·511 |
| | 45 | 1·473 | 1·415 | 1·238 | ·944 | ·531 |
| 7 | | 1·528 | 1·467 | 1·284 | ·979 | ·551 |
| | 15 | 1·582 | 1·520 | 1·330 | 1·014 | ·570 |
| | 30 | 1·637 | 1·572 | 1·375 | 1·048 | ·590 |
| | 45 | 1·692 | 1·624 | 1·421 | 1·083 | ·610 |
| 8 | | 1·746 | 1·677 | 1·467 | 1·118 | ·629 |
| | 15 | 1·801 | 1·729 | 1·513 | 1·153 | ·649 |
| | 30 | 1·855 | 1·782 | 1·559 | 1·188 | ·669 |
| | 45 | 1·910 | 1·834 | 1·605 | 1·223 | ·689 |
| 9 | | 1·965 | 1·886 | 1·651 | 1·258 | ·708 |
| | 15 | 2·019 | 1·939 | 1·696 | 1·293 | ·728 |
| | 30 | 2·074 | 1·991 | 1·742 | 1·328 | ·748 |
| | 45 | 2·128 | 2·044 | 1·788 | 1·363 | ·767 |
| 10 | | 2·183 | 2·096 | 1·834 | 1·398 | ·787 |
| | 15 | 2·238 | 2·148 | 1·880 | 1·433 | ·807 |
| | 30 | 2·292 | 2·201 | 1·926 | 1·468 | ·827 |
| | 45 | 2·347 | 2·254 | 1·972 | 1·503 | ·846 |
| 11 | | 2·401 | 2·306 | 2·018 | 1·538 | ·866 |
| | 15 | 2·456 | 2·359 | 2·064 | 1·574 | ·886 |
| | 30 | 2·511 | 2·411 | 2·110 | 1·609 | ·906 |
| | 45 | 2·566 | 2·464 | 2·156 | 1·644 | ·926 |
| 12 | | 2·620 | 2·516 | 2·203 | 1·680 | ·946 |
| | 15 | 2·675 | 2·569 | 2·249 | 1·715 | ·966 |
| | 30 | 2·730 | 2·621 | 2.295 | 1·750 | ·985 |

# TABLE OF ORDINATES.—Continued.

Ordinates 10 feet apart.—Chord 100 feet.

| Deflexion Angle in Degrees and Minutes. | | Distances of the Ordinates from the end of the 100 feet Chord. | | | | |
|---|---|---|---|---|---|---|
| | | 50 feet. | 40 feet. | 30 feet. | 20 feet. | 10 feet. |
| | | Lengths of Ordinates in feet. | | | | |
| ° | ′ | | | | | |
| 12 | 45 | 2·785 | 2·674 | 2·341 | 1·785 | 1·005 |
| 13 | | 2·839 | 2·726 | 2·387 | 1·820 | 1·025 |
| | 15 | 2·894 | 2·779 | 2·433 | 1·855 | 1·045 |
| | 30 | 2·949 | 2·832 | 2·479 | 1·891 | 1·065 |
| | 45 | 3·000 | 2·884 | 2·525 | 1·926 | 1·085 |
| 14° | | 3·058 | 2·937 | 2·571 | 1·961 | 1·105 |
| | 15 | 3·113 | 2·989 | 2·618 | 1·996 | 1·124 |
| | 30 | 3·168 | 3·042 | 2·664 | 2·031 | 1·144 |
| | 45 | 3·222 | 3·094 | 2·710 | 2·067 | 1·164 |
| 15° | | 3·277 | 3·147 | 2·756 | 2·102 | 1·184 |
| | 15 | 3·332 | 3·200 | 2·802 | 2·137 | 1·204 |
| | 30 | 3·387 | 3·252 | 2·848 | 2·172 | 1·224 |
| | 45 | 3·442 | 3·305 | 2·895 | 2·208 | 1·244 |
| 16° | | 3·496 | 3·358 | 2·941 | 2·243 | 1·264 |
| | 30 | 3·606 | 3·463 | 3·033 | 2·314 | 1·304 |
| 17° | | 3·716 | 3·569 | 3·125 | 2·384 | 1·344 |
| | 30 | 3·826 | 3·674 | 3·218 | 2·455 | 1·384 |
| 18° | | 3·935 | 3·779 | 3·310 | 2·525 | 1·424 |
| | 30 | 4·045 | 3·885 | 3·403 | 2·596 | 1·464 |
| 19° | | 4·155 | 3·990 | 3·495 | 2·666 | 1·504 |
| | 30 | 4·265 | 4·096 | 3·588 | 2·737 | 1·544 |
| 20° | | 4·375 | 4·201 | 3·680 | 2·808 | 1·583 |

# A TABLE OF THE SQUARES AND SQUARE ROOTS OF NUMBERS.

From 1 to 1000.

| No. | Squares. | Square Roots. | No. | Squares. | Square Roots. |
|---|---|---|---|---|---|
| 1 | 1 | 1·0000 | 44 | 1936 | 6·6332 |
| 2 | 4 | 1·4142 | 45 | 2025 | 6·7082 |
| 3 | 9 | 1·7320 | 46 | 2116 | 6·7823 |
| 4 | 16 | 2·0000 | 47 | 2209 | 6·8556 |
| 5 | 25 | 2·2360 | 48 | 2304 | 6·9282 |
| 6 | 36 | 2·4495 | 49 | 2401 | 7·0000 |
| 7 | 49 | 2·6457 | 50 | 2500 | 7·0711 |
| 8 | 64 | 2·8284 | 51 | 2601 | 7·1414 |
| 9 | 81 | 3·0000 | 52 | 2704 | 7·2111 |
| 10 | 100 | 3·1623 | 53 | 2809 | 7·2801 |
| 11 | 121 | 3·3166 | 54 | 2916 | 7·3485 |
| 12 | 144 | 3·4641 | 55 | 3025 | 7·4162 |
| 13 | 169 | 3·6055 | 56 | 3136 | 7·4833 |
| 14 | 196 | 3·7416 | 57 | 3249 | 7·5498 |
| 15 | 225 | 3·8730 | 58 | 3364 | 7·6158 |
| 16 | 256 | 4·0000 | 59 | 3481 | 7·6811 |
| 17 | 289 | 4·1231 | 60 | 3600 | 7·7460 |
| 18 | 324 | 4·2426 | 61 | 3721 | 7·8102 |
| 19 | 361 | 4·3589 | 62 | 3844 | 7·8740 |
| 20 | 400 | 4·4721 | 63 | 3969 | 7·9372 |
| 21 | 441 | 4·5826 | 64 | 4096 | 8·0000 |
| 22 | 484 | 4·6904 | 65 | 4225 | 8·0623 |
| 23 | 529 | 4·7958 | 66 | 4356 | 8·1240 |
| 24 | 576 | 4·8990 | 67 | 4489 | 8·1853 |
| 25 | 625 | 5·0000 | 68 | 4624 | 8·2462 |
| 26 | 676 | 5·0990 | 69 | 4761 | 8·3066 |
| 27 | 729 | 5·1961 | 70 | 4900 | 8·3666 |
| 28 | 784 | 5·2915 | 71 | 5041 | 8·4261 |
| 29 | 841 | 5·3852 | 72 | 5184 | 8·4853 |
| 30 | 900 | 5·4772 | 73 | 5329 | 8·5440 |
| 31 | 961 | 5·5678 | 74 | 5476 | 8·6023 |
| 32 | 1024 | 5·6568 | 75 | 5625 | 8·6603 |
| 33 | 1089 | 5·7446 | 76 | 5776 | 8·7178 |
| 34 | 1156 | 5·8309 | 77 | 5929 | 8·7750 |
| 35 | 1225 | 5·9161 | 78 | 6084 | 8·8318 |
| 36 | 1296 | 6·0000 | 79 | 6241 | 8·8882 |
| 37 | 1369 | 6·0828 | 80 | 6400 | 8·9443 |
| 38 | 1444 | 6·1644 | 81 | 6561 | 9·0000 |
| 39 | 1521 | 6·2450 | 82 | 6724 | 9·0554 |
| 40 | 1600 | 6·3246 | 83 | 6889 | 9·1104 |
| 41 | 1681 | 6·4031 | 84 | 7056 | 9·1651 |
| 42 | 1764 | 6·4807 | 85 | 7225 | 9·2195 |
| 43 | 1849 | 6·5574 | 86 | 7396 | 9·2736 |

## A TABLE OF THE SQUARES AND SQUARE ROOTS OF NUMBERS.—CONTINUED.

From 1 to 1000.

| No. | Squares. | Square Roots. | No. | Squares. | Square Roots. |
|---|---|---|---|---|---|
| 87 | 7569 | 9·3274 | 130 | 16900 | 11·4017 |
| 88 | 7744 | 9·3808 | 131 | 17161 | 11·4455 |
| 89 | 7921 | 9·4340 | 132 | 17424 | 11·4891 |
| 90 | 8100 | 9·4868 | 133 | 17689 | 11·5326 |
| 91 | 8281 | 9·5394 | 134 | 17956 | 11·5758 |
| 92 | 8464 | 9·5917 | 135 | 18225 | 11·6189 |
| 93 | 8649 | 9·6436 | 136 | 18496 | 11·6619 |
| 94 | 8836 | 9·6954 | 137 | 18769 | 11·7047 |
| 95 | 9025 | 9·7468 | 138 | 19044 | 11·7473 |
| 96 | 9216 | 9·7979 | 139 | 19321 | 11·7898 |
| 97 | 9409 | 9·8488 | 140 | 19600 | 11·8322 |
| 98 | 9604 | 9·8995 | 141 | 19881 | 11·8743 |
| 99 | 9801 | 9·9499 | 142 | 20164 | 11·9164 |
| 100 | 10000 | 10·0000 | 143 | 20449 | 11·9583 |
| 101 | 10201 | 10·0499 | 144 | 20736 | 12·0000 |
| 102 | 10404 | 10·0995 | 145 | 21025 | 12·0416 |
| 103 | 10609 | 10·1489 | 146 | 21316 | 12·0830 |
| 104 | 10816 | 10·1980 | 147 | 21609 | 12·1244 |
| 105 | 11025 | 10·2469 | 148 | 21904 | 12·1655 |
| 106 | 11236 | 10·2956 | 149 | 22201 | 12·2065 |
| 107 | 11449 | 10·3440 | 150 | 22500 | 12·2474 |
| 108 | 11664 | 10·3923 | 151 | 22801 | 12·2882 |
| 109 | 11881 | 10·4403 | 152 | 23104 | 12·3288 |
| 110 | 12100 | 10·4881 | 153 | 23409 | 12·3693 |
| 111 | 12321 | 10·5356 | 154 | 23716 | 12·4097 |
| 112 | 12544 | 10·5830 | 155 | 24025 | 12·4499 |
| 113 | 12769 | 10·6301 | 156 | 24336 | 12·4900 |
| 114 | 12996 | 10·6771 | 157 | 24649 | 12·5300 |
| 115 | 13225 | 10·7238 | 158 | 24964 | 12·5700 |
| 116 | 13456 | 10·7703 | 159 | 25281 | 12·6095 |
| 117 | 13689 | 10·8166 | 160 | 25600 | 12·6491 |
| 118 | 13924 | 10·8628 | 161 | 25921 | 12·6886 |
| 119 | 14161 | 10·9087 | 162 | 26244 | 12·7279 |
| 120 | 14400 | 10·9544 | 163 | 26569 | 12·7671 |
| 121 | 14641 | 11·0000 | 164 | 26896 | 12·8062 |
| 122 | 14884 | 11·0454 | 165 | 27225 | 12·8452 |
| 123 | 15129 | 11·0905 | 166 | 27556 | 12·8841 |
| 124 | 15376 | 11·1355 | 167 | 27889 | 12·9228 |
| 125 | 15625 | 11·1803 | 168 | 28224 | 12·9615 |
| 126 | 15876 | 11.2250 | 169 | 28561 | 13·0000 |
| 127 | 16129 | 11·2694 | 170 | 28900 | 13·0384 |
| 128 | 16384 | 11·3137 | 171 | 29241 | 13·0767 |
| 129 | 16641 | 11·3578 | 172 | 29584 | 13·1149 |

# A TABLE OF THE SQUARES AND SQUARE ROOTS OF NUMBERS.—Continued.

From 1 to 1000.

| No. | Squares. | Square Roots. | No. | Squares. | Square Roots. |
|---|---|---|---|---|---|
| 173 | 29929 | 13·1529 | 216 | 46656 | 14·6969 |
| 174 | 30276 | 13·1909 | 217 | 47089 | 14·7309 |
| 175 | 30625 | 13·2287 | 218 | 47524 | 14·7648 |
| 176 | 30976 | 13·2665 | 219 | 47961 | 14·7986 |
| 177 | 31329 | 13·3041 | 220 | 48400 | 14·8324 |
| 178 | 31684 | 13·3417 | 221 | 48841 | 14·8661 |
| 179 | 32041 | 13·3791 | 222 | 49284 | 14·8997 |
| 180 | 32400 | 13·4164 | 223 | 49729 | 14·9332 |
| 181 | 32761 | 13·4536 | 224 | 50176 | 14·9666 |
| 182 | 33124 | 13·4907 | 225 | 50625 | 15·0000 |
| 183 | 33489 | 13·5277 | 226 | 51076 | 15·0333 |
| 184 | 33856 | 13·5647 | 227 | 51529 | 15·0665 |
| 185 | 34225 | 13·6015 | 228 | 51984 | 15·0997 |
| 186 | 34596 | 13·6382 | 229 | 52441 | 15·1327 |
| 187 | 34969 | 13·6748 | 230 | 52900 | 15·1657 |
| 188 | 35344 | 13·7113 | 231 | 53361 | 15·1987 |
| 189 | 35721 | 13·7477 | 232 | 53824 | 15·2315 |
| 190 | 36100 | 13·7840 | 233 | 54289 | 15·2643 |
| 191 | 36481 | 13·8203 | 234 | 54756 | 15·2970 |
| 192 | 36864 | 13·8564 | 235 | 55225 | 15·3297 |
| 193 | 37249 | 13·8924 | 236 | 55696 | 15·3623 |
| 194 | 37636 | 13·9284 | 237 | 56169 | 15·3948 |
| 195 | 38025 | 13·9642 | 238 | 56644 | 15·4272 |
| 196 | 38416 | 14·0000 | 239 | 57121 | 15·4596 |
| 197 | 38809 | 14·0357 | 240 | 57600 | 15·4919 |
| 198 | 39204 | 14·0712 | 241 | 58081 | 15·5242 |
| 199 | 39601 | 14·1067 | 242 | 58564 | 15·5563 |
| 200 | 40000 | 14·1421 | 243 | 59049 | 15·5885 |
| 201 | 40401 | 14·1774 | 244 | 59536 | 15·6205 |
| 202 | 40804 | 14·2127 | 245 | 60025 | 15·6525 |
| 203 | 41209 | 14·2478 | 246 | 60516 | 15·6844 |
| 204 | 41616 | 14·2828 | 247 | 61009 | 15·7162 |
| 205 | 42025 | 14·3178 | 248 | 61504 | 15·7480 |
| 206 | 42436 | 14·3527 | 249 | 62001 | 15·7797 |
| 207 | 42849 | 14·3874 | 250 | 62500 | 15·8114 |
| 208 | 43264 | 14·4222 | 251 | 63001 | 15·8430 |
| 209 | 43681 | 14·4568 | 252 | 63504 | 15·8745 |
| 210 | 44100 | 14·4914 | 253 | 64009 | 15·9060 |
| 211 | 44521 | 14·5258 | 254 | 64516 | 15·9374 |
| 212 | 44944 | 14·5602 | 255 | 65025 | 15·9687 |
| 213 | 45369 | 14·5945 | 256 | 65536 | 16·0000 |
| 214 | 45796 | 14·6287 | 257 | 66049 | 16·0312 |
| 215 | 46225 | 14·6629 | 258 | 66564 | 16·0624 |

## A TABLE OF THE SQUARES AND SQUARE ROOTS OF NUMBERS.—Continued.

From 1 to 1000.

| No. | Squares. | Square Roots. | No. | Squares. | Square Roots. |
|---|---|---|---|---|---|
| 259 | 67081 | 16.0935 | 302 | 91204 | 17·3781 |
| 260 | 67600 | 16·1245 | 303 | 91809 | 17·4069 |
| 261 | 68121 | 16·1555 | 304 | 92416 | 17·4356 |
| 262 | 68644 | 16·1864 | 305 | 93025 | 17·4642 |
| 263 | 69169 | 16·2173 | 306 | 93636 | 17·4928 |
| 264 | 69696 | 16·2481 | 307 | 94249 | 17·5214 |
| 265 | 70225 | 16·2788 | 308 | 94864 | 17·5499 |
| 266 | 70756 | 16·3095 | 309 | 95481 | 17·5784 |
| 267 | 71289 | 16·3401 | 310 | 96100 | 17·6068 |
| 268 | 71824 | 16·3707 | 311 | 96721 | 17·6352 |
| 269 | 72361 | 16·4012 | 312 | 97344 | 17·6635 |
| 270 | 72900 | 16·4317 | 313 | 97969 | 17·6918 |
| 271 | 73441 | 16·4621 | 314 | 98596 | 17·7200 |
| 272 | 73984 | 16·4924 | 315 | 99225 | 17·7482 |
| 273 | 74529 | 16·5227 | 316 | 99856 | 17·7764 |
| 274 | 75076 | 16·5529 | 317 | 100489 | 17·8045 |
| 275 | 75625 | 16·5831 | 318 | 101124 | 17·8325 |
| 276 | 76176 | 16·6132 | 319 | 101761 | 17·8606 |
| 277 | 76729 | 16·6433 | 320 | 102400 | 17·8885 |
| 278 | 77284 | 16·6733 | 321 | 103041 | 17·9165 |
| 279 | 77841 | 16·7033 | 322 | 103684 | 17·9444 |
| 280 | 78400 | 16·7332 | 323 | 104329 | 17·9722 |
| 281 | 78961 | 16·7630 | 324 | 104976 | 18·0000 |
| 282 | 79524 | 16·7928 | 325 | 105625 | 18·0277 |
| 283 | 80089 | 16·8226 | 326 | 106276 | 18·0555 |
| 284 | 80656 | 16·8523 | 327 | 106929 | 18·0831 |
| 285 | 81225 | 16·8819 | 328 | 107584 | 18·1108 |
| 286 | 81796 | 16·9115 | 329 | 108241 | 18·1384 |
| 287 | 82369 | 16·9411 | 330 | 108900 | 18·1659 |
| 288 | 82944 | 16·9706 | 331 | 109561 | 18·1934 |
| 289 | 83521 | 17·0000 | 332 | 110224 | 18·2209 |
| 290 | 84100 | 17·0294 | 333 | 110889 | 18·2483 |
| 291 | 84681 | 17·0587 | 334 | 111556 | 18·2757 |
| 292 | 85264 | 17·0880 | 335 | 112225 | 18·3030 |
| 293 | 85849 | 17·1172 | 336 | 112896 | 18·3303 |
| 294 | 86436 | 17·1464 | 337 | 113569 | 18·3576 |
| 295 | 87025 | 17·1756 | 338 | 114244 | 18·3848 |
| 296 | 87616 | 17·2046 | 339 | 114921 | 18·4119 |
| 297 | 88209 | 17·2337 | 340 | 115600 | 18·4391 |
| 298 | 88804 | 17·2627 | 341 | 116281 | 18·4662 |
| 299 | 89401 | 17·2916 | 342 | 116964 | 18·4932 |
| 300 | 90000 | 17·3205 | 343 | 117649 | 18·5203 |
| 301 | 90601 | 17·3493 | 344 | 118336 | 18·5472 |

# A TABLE OF THE SQUARES AND SQUARE ROOTS OF NUMBERS.—Continued.

From 1 to 1000.

| No. | Squares. | Square Roots. | No. | Squares. | Square Roots. |
|---|---|---|---|---|---|
| 345 | 119025 | 18·5742 | 388 | 150544 | 19·6977 |
| 346 | 119716 | 18·6011 | 389 | 151321 | 19·7231 |
| 347 | 120409 | 18·6279 | 390 | 152100 | 19·7484 |
| 348 | 121104 | 18·6548 | 391 | 152881 | 19·7737 |
| 349 | 121801 | 18·6815 | 392 | 153664 | 19·7990 |
| 350 | 122500 | 18·7083 | 393 | 154449 | 19·8242 |
| 351 | 123201 | 18·7350 | 394 | 155236 | 19·8494 |
| 352 | 123904 | 18·7617 | 395 | 156025 | 19·8746 |
| 353 | 124609 | 18·7883 | 396 | 156816 | 19·8997 |
| 354 | 125316 | 18·8149 | 397 | 157609 | 19·9248 |
| 355 | 126025 | 18·8414 | 398 | 158404 | 19·9499 |
| 356 | 126736 | 18·8680 | 399 | 159201 | 19·9750 |
| 357 | 127449 | 18·8944 | 400 | 160000 | 20·0000 |
| 358 | 128164 | 18·9209 | 401 | 160801 | 20·0250 |
| 359 | 128881 | 18·9473 | 402 | 161604 | 20·0499 |
| 360 | 129600 | 18·9737 | 403 | 162409 | 20·0749 |
| 361 | 130321 | 19·0000 | 404 | 163216 | 20·0997 |
| 362 | 131044 | 19·0263 | 405 | 164025 | 20·1246 |
| 363 | 131769 | 19·0526 | 406 | 164836 | 20·1494 |
| 364 | 132496 | 19·0788 | 407 | 165649 | 20·1742 |
| 365 | 133225 | 19·1050 | 408 | 166464 | 20·1990 |
| 366 | 133956 | 19·1311 | 409 | 167281 | 20·2237 |
| 367 | 134689 | 19·1572 | 410 | 168100 | 20·2485 |
| 368 | 135424 | 19·1833 | 411 | 168921 | 20·2731 |
| 369 | 136161 | 19·2094 | 412 | 169744 | 20·2978 |
| 370 | 136900 | 19·2354 | 413 | 170569 | 20·3224 |
| 371 | 137641 | 19·2614 | 414 | 171396 | 20·3470 |
| 372 | 138384 | 19·2873 | 415 | 172225 | 20·3715 |
| 373 | 139129 | 19·3132 | 416 | 173056 | 20·3961 |
| 374 | 139876 | 19·3391 | 417 | 173889 | 20·4206 |
| 375 | 140625 | 19·3649 | 418 | 174724 | 20·4450 |
| 376 | 141376 | 19·3907 | 419 | 175561 | 20·4695 |
| 377 | 142129 | 19·4165 | 420 | 176400 | 20·4939 |
| 378 | 142884 | 19·4422 | 421 | 177241 | 20·5183 |
| 379 | 143641 | 19·4679 | 422 | 178084 | 20·5426 |
| 380 | 144400 | 19·4936 | 423 | 178929 | 20·5670 |
| 381 | 145161 | 19·5192 | 424 | 179776 | 20·5913 |
| 382 | 145924 | 19·5448 | 425 | 180625 | 20·6155 |
| 383 | 146689 | 19·5704 | 426 | 181476 | 20·6398 |
| 384 | 147456 | 19·5959 | 427 | 182329 | 20·6640 |
| 385 | 148225 | 19·6214 | 428 | 183184 | 20·6882 |
| 386 | 148996 | 19·6469 | 429 | 184041 | 20·7123 |
| 387 | 149769 | 19·6723 | 430 | 184900 | 20·7364 |

# A TABLE OF THE SQUARES AND SQUARE ROOTS OF NUMBERS.—Continued.

From 1 to 1000.

| No. | Squares. | Square Roots. | No. | Squares. | Square Roots. |
|---|---|---|---|---|---|
| 431 | 185761 | 20·7605 | 474 | 224676 | 21·7715 |
| 432 | 186624 | 20·7846 | 475 | 225625 | 21·7945 |
| 433 | 187489 | 20·8086 | 476 | 226576 | 21·8174 |
| 434 | 188356 | 20·8327 | 477 | 227529 | 21·8403 |
| 435 | 189225 | 20·8566 | 478 | 228484 | 21·8632 |
| 436 | 190096 | 20·8806 | 479 | 229441 | 21·8861 |
| 437 | 190969 | 20·9045 | 480 | 230400 | 21·9089 |
| 438 | 191844 | 20·9284 | 481 | 231361 | 21·9317 |
| 439 | 192721 | 20·9523 | 482 | 232324 | 21·9545 |
| 440 | 193600 | 20·9762 | 483 | 233289 | 21·9773 |
| 441 | 194481 | 21·0000 | 484 | 234256 | 22·0000 |
| 442 | 195364 | 21·0238 | 485 | 235225 | 22·0227 |
| 443 | 196249 | 21·0476 | 486 | 236196 | 22·0454 |
| 444 | 197136 | 21·0713 | 487 | 237169 | 22·0689 |
| 445 | 198025 | 21·0950 | 488 | 238144 | 22·0907 |
| 446 | 198916 | 21·1187 | 489 | 239121 | 22·1133 |
| 447 | 199809 | 21·1424 | 490 | 240100 | 22·1359 |
| 448 | 200704 | 21·1660 | 491 | 241081 | 22·1585 |
| 449 | 201601 | 21·1896 | 492 | 242064 | 22·1811 |
| 450 | 202500 | 21·2132 | 493 | 243049 | 22·2036 |
| 451 | 203401 | 21·2368 | 494 | 244036 | 22·2261 |
| 452 | 204304 | 21·2603 | 495 | 245025 | 22·2486 |
| 453 | 205209 | 21·2838 | 496 | 246016 | 22·2711 |
| 454 | 206116 | 21·3073 | 497 | 247009 | 22·2935 |
| 455 | 207025 | 21·3307 | 498 | 248004 | 22·3159 |
| 456 | 207936 | 21·3542 | 499 | 249001 | 22·3383 |
| 457 | 208849 | 21·3776 | 500 | 250000 | 22·3607 |
| 458 | 209764 | 21·4009 | 501 | 251001 | 22·3830 |
| 459 | 210681 | 21·4243 | 502 | 252004 | 22·4054 |
| 460 | 211600 | 21·4476 | 503 | 253009 | 22·4277 |
| 461 | 212521 | 21·4709 | 504 | 254016 | 22·4499 |
| 462 | 213444 | 21·4942 | 505 | 255025 | 22·4722 |
| 463 | 214369 | 21·5174 | 506 | 256036 | 22·4944 |
| 464 | 215296 | 21·5407 | 507 | 257049 | 22·5167 |
| 465 | 216225 | 21·5639 | 508 | 258064 | 22·5388 |
| 466 | 217156 | 21·5870 | 509 | 259081 | 22·5610 |
| 467 | 218089 | 21·6102 | 510 | 260100 | 22·5832 |
| 468 | 219024 | 21·6333 | 511 | 261121 | 22·6053 |
| 469 | 219961 | 21·6564 | 512 | 262144 | 22·6274 |
| 470 | 220900 | 21·6795 | 513 | 263169 | 22·6495 |
| 471 | 221841 | 21·7025 | 514 | 264196 | 22·6716 |
| 472 | 222784 | 21·7256 | 515 | 265225 | 22·6936 |
| 473 | 223729 | 21·7486 | 516 | 266256 | 22·7156 |

# A TABLE OF THE SQUARES AND SQUARE ROOTS OF NUMBERS.—CONTINUED.

From 1 to 1000.

| No. | Squares. | Square Roots. | No. | Squares. | Square Roots. |
|---|---|---|---|---|---|
| 517 | 267289 | 22·7376 | 560 | 313600 | 23·6643 |
| 518 | 268324 | 22·7596 | 561 | 314721 | 23·6854 |
| 519 | 269361 | 22·7816 | 562 | 315844 | 23·7065 |
| 520 | 270400 | 22·8035 | 563 | 316969 | 23·7276 |
| 521 | 271441 | 22·8254 | 564 | 318096 | 23·7487 |
| 522 | 272484 | 22·8473 | 565 | 319225 | 23·7697 |
| 523 | 273529 | 22·8692 | 566 | 320356 | 23·7907 |
| 524 | 274576 | 22·8910 | 567 | 321489 | 23·8118 |
| 525 | 275625 | 22·9129 | 568 | 322624 | 23·8327 |
| 526 | 276676 | 22·9347 | 569 | 323761 | 23·8537 |
| 527 | 277729 | 22·9565 | 570 | 324900 | 23·8747 |
| 528 | 278784 | 22·9782 | 571 | 326041 | 23·8956 |
| 529 | 279841 | 23·0000 | 572 | 327184 | 23·9165 |
| 530 | 280900 | 23·0217 | 573 | 328329 | 23·9374 |
| 531 | 281961 | 23·0434 | 574 | 329476 | 23·9583 |
| 532 | 283024 | 23·0651 | 575 | 330625 | 23·9792 |
| 533 | 284089 | 23·0868 | 576 | 331776 | 24·0000 |
| 534 | 285156 | 23·1084 | 577 | 332929 | 24·0208 |
| 535 | 286225 | 23·1301 | 578 | 334084 | 24·0416 |
| 536 | 287296 | 23·1517 | 579 | 335241 | 24·0624 |
| 537 | 288369 | 23·1733 | 580 | 336400 | 24·0832 |
| 538 | 289444 | 23·1948 | 581 | 337561 | 24·1039 |
| 539 | 290521 | 23·2164 | 582 | 338724 | 24·1247 |
| 540 | 291600 | 23·2379 | 583 | 339889 | 24·1454 |
| 541 | 292681 | 23·2594 | 584 | 341056 | 24·1661 |
| 542 | 293764 | 23·2809 | 585 | 342225 | 24·1868 |
| 543 | 294849 | 23·3021 | 586 | 343396 | 24·2074 |
| 544 | 295936 | 23·3238 | 587 | 344569 | 24·2281 |
| 545 | 297025 | 23·3452 | 588 | 345744 | 24·2487 |
| 546 | 298116 | 23·3666 | 589 | 346921 | 24·2693 |
| 547 | 299209 | 23·3880 | 590 | 348100 | 24·2899 |
| 548 | 300304 | 23·4094 | 591 | 349281 | 24·3105 |
| 549 | 301401 | 23·4307 | 592 | 350464 | 24·3310 |
| 550 | 302500 | 23·4521 | 593 | 351649 | 24·3516 |
| 551 | 303601 | 23·4734 | 594 | 352836 | 24·3721 |
| 552 | 304704 | 23·4947 | 595 | 354025 | 24·3926 |
| 553 | 305809 | 23·5159 | 596 | 355216 | 24·4131 |
| 554 | 306916 | 23·5372 | 597 | 356409 | 24·4336 |
| 555 | 308025 | 23·5584 | 598 | 357604 | 24·4540 |
| 556 | 309136 | 23·5796 | 599 | 358801 | 24·4745 |
| 557 | 310249 | 23·6008 | 600 | 360000 | 24·4949 |
| 558 | 311364 | 23·6220 | 601 | 361201 | 24·5153 |
| 559 | 312481 | 23·6432 | 602 | 362404 | 24·5357 |

# A TABLE OF THE SQUARES AND SQUARE ROOTS OF NUMBERS.—Continued.

From 1 to 1000.

| No. | Squares. | Square Roots. | No. | Squares. | Square Roots. |
|---|---|---|---|---|---|
| 603 | 363609 | 24·5560 | 646 | 417316 | 25·4165 |
| 604 | 364816 | 24·5764 | 647 | 418609 | 25·4362 |
| 605 | 366025 | 24·5967 | 648 | 419904 | 25·4558 |
| 606 | 367236 | 24·6171 | 649 | 421201 | 25·4755 |
| 607 | 368449 | 24·6374 | 650 | 422500 | 25·4950 |
| 608 | 369664 | 24·6576 | 651 | 423801 | 25·5147 |
| 609 | 370881 | 24·6779 | 652 | 425104 | 25·5343 |
| 610 | 372100 | 24·6982 | 653 | 426409 | 25·5539 |
| 611 | 373321 | 24·7184 | 654 | 427716 | 25·5734 |
| 612 | 374544 | 24·7386 | 655 | 429025 | 25·5930 |
| 613 | 375769 | 24·7588 | 656 | 430336 | 25·6125 |
| 614 | 376996 | 24·7790 | 657 | 431649 | 25·6320 |
| 615 | 378225 | 24·7992 | 658 | 432964 | 25·6515 |
| 616 | 379456 | 24·8193 | 659 | 434281 | 25·6710 |
| 617 | 380689 | 24·8395 | 660 | 435600 | 25·6905 |
| 618 | 381924 | 24·8596 | 661 | 436921 | 25·7099 |
| 619 | 383161 | 24·8797 | 662 | 438244 | 25·7204 |
| 620 | 384400 | 24·8998 | 663 | 439569 | 25·7488 |
| 621 | 385641 | 24·9199 | 664 | 440896 | 25·7682 |
| 622 | 386884 | 24·9399 | 665 | 442225 | 25·7876 |
| 623 | 388129 | 24·9600 | 666 | 443556 | 25·8070 |
| 624 | 389376 | 24·9800 | 667 | 444889 | 25·8263 |
| 625 | 390625 | 25·0000 | 668 | 446224 | 25·8457 |
| 626 | 391876 | 25·0200 | 669 | 447561 | 25·8650 |
| 627 | 393129 | 25·0400 | 670 | 448900 | 25·8844 |
| 628 | 394384 | 25·0600 | 671 | 450241 | 25·9037 |
| 629 | 395641 | 25·0799 | 672 | 451584 | 25·9230 |
| 630 | 396900 | 25·0998 | 673 | 452929 | 25·9422 |
| 631 | 398161 | 25·1197 | 674 | 454276 | 25·9615 |
| 632 | 399424 | 25·1396 | 675 | 455625 | 25·9808 |
| 633 | 400689 | 25·1595 | 676 | 456976 | 26·0000 |
| 634 | 401956 | 25·1794 | 677 | 458329 | 26·0192 |
| 635 | 403225 | 25·1992 | 678 | 459684 | 26·0384 |
| 636 | 404496 | 25·2190 | 679 | 461041 | 26·0576 |
| 637 | 405769 | 25·2389 | 680 | 462400 | 26·0768 |
| 638 | 407044 | 25·2587 | 681 | 463761 | 26·0960 |
| 639 | 408321 | 25·2785 | 682 | 465124 | 26·1151 |
| 640 | 409600 | 25·2982 | 683 | 466489 | 26·1343 |
| 641 | 410881 | 25·3180 | 684 | 467856 | 26·1534 |
| 642 | 412164 | 25·3377 | 685 | 469225 | 26·1725 |
| 643 | 413449 | 25·3574 | 686 | 470596 | 26·1916 |
| 644 | 414736 | 25·3772 | 687 | 471969 | 26·2107 |
| 645 | 416025 | 25·3968 | 688 | 473344 | 26·2297 |

## A TABLE OF THE SQUARES AND SQUARE ROOTS OF NUMBERS.—Continued.

From 1 to 1000.

| No. | Squares. | Square Roots. | No. | Squares. | Square Roots. |
|---|---|---|---|---|---|
| 689 | 474721 | 26·2488 | 732 | 535824 | 27·0555 |
| 690 | 476100 | 26·2678 | 733 | 537289 | 27·0740 |
| 691 | 477481 | 26·2869 | 734 | 538756 | 27·0924 |
| 692 | 478864 | 26·3059 | 735 | 540225 | 27·1109 |
| 693 | 480249 | 26·3249 | 736 | 541696 | 27·1293 |
| 694 | 481636 | 26·3439 | 737 | 543169 | 27·1477 |
| 695 | 483025 | 26·3629 | 738 | 544644 | 27·1662 |
| 696 | 484416 | 26·3818 | 739 | 546121 | 27·1846 |
| 697 | 485809 | 26·4008 | 740 | 547600 | 27·2029 |
| 698 | 487204 | 26·4197 | 741 | 549081 | 27·2213 |
| 699 | 488601 | 26·4386 | 742 | 550564 | 27·2397 |
| 700 | 490000 | 26·4575 | 743 | 552049 | 27·2580 |
| 701 | 491401 | 26·4764 | 744 | 553536 | 27·2764 |
| 702 | 492804 | 26·4953 | 745 | 555025 | 27·2947 |
| 703 | 494209 | 26·5141 | 746 | 556516 | 27·3130 |
| 704 | 495616 | 26·5330 | 747 | 558009 | 27·3313 |
| 705 | 497025 | 26·5518 | 748 | 559504 | 27·3496 |
| 706 | 498436 | 26·5707 | 749 | 561001 | 27·3679 |
| 707 | 499849 | 26·5895 | 750 | 562500 | 27·3861 |
| 708 | 501264 | 26·6083 | 751 | 564001 | 27·4044 |
| 709 | 502681 | 26·6271 | 752 | 565504 | 27·4226 |
| 710 | 504100 | 26·6458 | 753 | 567009 | 27·4408 |
| 711 | 505521 | 26·6646 | 754 | 568516 | 27·4591 |
| 712 | 506944 | 26·6833 | 755 | 570025 | 27·4773 |
| 713 | 508369 | 26·7021 | 756 | 571536 | 27·4955 |
| 714 | 509796 | 26·7208 | 757 | 573049 | 27·5136 |
| 715 | 511225 | 26·7395 | 758 | 574564 | 27·5318 |
| 716 | 512656 | 26·7582 | 759 | 576081 | 27·5500 |
| 717 | 514089 | 26·7769 | 760 | 577600 | 27·5681 |
| 718 | 515524 | 26·7955 | 761 | 579121 | 27·5862 |
| 719 | 516961 | 26·8142 | 762 | 580644 | 27·6043 |
| 720 | 518400 | 26·8328 | 763 | 582169 | 27·6225 |
| 721 | 519841 | 26·8514 | 764 | 583696 | 27·6405 |
| 722 | 521284 | 26·8701 | 765 | 585225 | 27·6586 |
| 723 | 522729 | 26·8887 | 766 | 586756 | 27·6767 |
| 724 | 524176 | 26·9072 | 767 | 588289 | 27·6948 |
| 725 | 525625 | 26·9258 | 768 | 589824 | 27·7128 |
| 726 | 527076 | 26·9444 | 769 | 591361 | 27·7308 |
| 727 | 528529 | 26·9629 | 770 | 592900 | 27·7489 |
| 728 | 529984 | 26·9815 | 771 | 594441 | 27·7669 |
| 729 | 531441 | 27·0000 | 772 | 595984 | 27·7849 |
| 730 | 532900 | 27·0185 | 773 | 597529 | 27·8029 |
| 731 | 534361 | 27·0370 | 774 | 599076 | 27·8209 |

## A TABLE OF THE SQUARES AND SQUARE ROOTS OF NUMBERS.—CONTINUED.

From 1 to 1000.

| No. | Squares. | Square Roots. | No. | Squares. | Square Roots. |
|---|---|---|---|---|---|
| 775 | 600625 | 27·8388 | 818 | 669124 | 28·6007 |
| 776 | 602176 | 27·8568 | 819 | 670761 | 28·6182 |
| 777 | 603729 | 27·8747 | 820 | 672400 | 28·6356 |
| 778 | 605284 | 27·8926 | 821 | 674041 | 28·6531 |
| 779 | 606841 | 27·9106 | 822 | 675684 | 28·6705 |
| 780 | 608400 | 27·9285 | 823 | 677329 | 28·6880 |
| 781 | 609961 | 27·9464 | 824 | 678976 | 28·7054 |
| 782 | 611524 | 27·9643 | 825 | 680625 | 28·7228 |
| 783 | 613089 | 27·9821 | 826 | 682276 | 28·7402 |
| 784 | 614656 | 28·0000 | 827 | 683929 | 28·7576 |
| 785 | 616225 | 28·0179 | 828 | 685584 | 28·7750 |
| 786 | 617796 | 28·0357 | 829 | 687241 | 28·7924 |
| 787 | 619369 | 28·0535 | 830 | 688900 | 28·8097 |
| 788 | 620944 | 28·0713 | 831 | 690561 | 28·8271 |
| 789 | 622521 | 28·0891 | 832 | 692224 | 28·8444 |
| 790 | 624100 | 28·1069 | 833 | 693889 | 28·8617 |
| 791 | 625681 | 28·1247 | 834 | 695556 | 28·8791 |
| 792 | 627264 | 28·1425 | 835 | 697225 | 28·8964 |
| 793 | 628849 | 28·1603 | 836 | 698896 | 28·9137 |
| 794 | 630436 | 28·1780 | 837 | 700569 | 28·9310 |
| 795 | 632025 | 28·1957 | 838 | 702244 | 28·9482 |
| 796 | 633616 | 28·2135 | 839 | 703921 | 28·9655 |
| 797 | 635209 | 28·2312 | 840 | 705600 | 28·9828 |
| 798 | 636804 | 28·2489 | 841 | 707281 | 29·0000 |
| 799 | 638401 | 28·2666 | 842 | 708964 | 29·0172 |
| 800 | 640000 | 28·2843 | 843 | 710649 | 29·0345 |
| 801 | 641601 | 28·3019 | 844 | 712336 | 29·0517 |
| 802 | 643204 | 28·3196 | 845 | 714025 | 29·0689 |
| 803 | 644809 | 28·3373 | 846 | 715716 | 29·0861 |
| 804 | 646416 | 28·3549 | 847 | 717409 | 29·1033 |
| 805 | 648025 | 28·3726 | 848 | 719104 | 29·1204 |
| 806 | 649636 | 28·3901 | 849 | 720801 | 29·1376 |
| 807 | 651249 | 28·4077 | 850 | 722500 | 29·1548 |
| 808 | 652864 | 28·4253 | 851 | 724201 | 29·1719 |
| 809 | 654481 | 28·4429 | 852 | 725904 | 29·1890 |
| 810 | 656100 | 28·4605 | 853 | 727609 | 29·2062 |
| 811 | 657721 | 28·4781 | 854 | 729316 | 29·2233 |
| 812 | 659344 | 28·4956 | 855 | 731025 | 29·2404 |
| 813 | 660969 | 28·5132 | 856 | 732736 | 29·2575 |
| 814 | 662596 | 28·5307 | 857 | 734449 | 29·2746 |
| 815 | 664225 | 28·5482 | 858 | 736164 | 29·2916 |
| 816 | 665856 | 28·5657 | 859 | 737881 | 29·3087 |
| 817 | 667489 | 28·5832 | 860 | 739600 | 29·3258 |

## A TABLE OF THE SQUARES AND SQUARE ROOTS OF NUMBERS.—Continued.

From 1 to 1000.

| No. | Squares. | Square Roots. | No. | Squares. | Square Roots. |
|---|---|---|---|---|---|
| 861 | 741321 | 29·3428 | 904 | 817216 | 30·0666 |
| 862 | 743044 | 29·3598 | 905 | 819025 | 30·0832 |
| 863 | 744769 | 29·3769 | 906 | 820836 | 30·0998 |
| 864 | 746496 | 29·3939 | 907 | 822649 | 30·1164 |
| 865 | 748225 | 29·4109 | 908 | 824464 | 30·1330 |
| 866 | 749956 | 29·4279 | 909 | 826281 | 30·1496 |
| 867 | 751689 | 29·4449 | 910 | 828100 | 30·1662 |
| 868 | 753424 | 29·4618 | 911 | 829921 | 30·1828 |
| 869 | 755161 | 29·4788 | 912 | 831744 | 30·1993 |
| 870 | 756900 | 29·4958 | 913 | 833569 | 30 2159 |
| 871 | 758641 | 29·5127 | 914 | 835396 | 30·2324 |
| 872 | 760384 | 29·5296 | 915 | 837225 | 30·2490 |
| 873 | 762129 | 29·5466 | 916 | 839056 | 30·2655 |
| 874 | 763876 | 29·5635 | 917 | 840889 | 30·2820 |
| 875 | 765625 | 29·5804 | 918 | 842724 | 30·2985 |
| 876 | 767376 | 29·5973 | 919 | 844561 | 30·3150 |
| 877 | 769129 | 29·6142 | 920 | 846400 | 30·3315 |
| 878 | 770884 | 29·6311 | 921 | 848241 | 30·3480 |
| 879 | 772641 | 29·6479 | 922 | 850084 | 30·3645 |
| 880 | 774400 | 29·6648 | 923 | 851929 | 30·3809 |
| 881 | 776161 | 29·6816 | 924 | 853776 | 30·3974 |
| 882 | 777924 | 29·6985 | 925 | 855625 | 30·4138 |
| 883 | 779689 | 29·7153 | 926 | 857476 | 30·4302 |
| 884 | 781456 | 29·7321 | 927 | 859329 | 30·4467 |
| 885 | 783225 | 29·7489 | 928 | 861184 | 30·4631 |
| 886 | 784996 | 29·7658 | 929 | 863041 | 30·4795 |
| 887 | 786769 | 29·7825 | 930 | 864900 | 30·4959 |
| 888 | 788544 | 29·7993 | 931 | 866761 | 30·5123 |
| 889 | 790321 | 29·8161 | 932 | 868624 | 30·5287 |
| 890 | 792100 | 29·8329 | 933 | 870489 | 30·5450 |
| 891 | 793881 | 29·8496 | 934 | 872356 | 30·5614 |
| 892 | 795664 | 29·8664 | 935 | 874225 | 30·5778 |
| 893 | 797449 | 29·8831 | 936 | 876096 | 30·5941 |
| 894 | 799236 | 29·8998 | 937 | 877969 | 30·6105 |
| 895 | 801025 | 29·9166 | 938 | 879844 | 30·6268 |
| 896 | 802816 | 29·9333 | 939 | 881721 | 30·6431 |
| 897 | 804609 | 29·9500 | 940 | 883600 | 30·6594 |
| 898 | 806404 | 29·9666 | 941 | 885481 | 30·6757 |
| 999 | 808201 | 29·9833 | 942 | 887364 | 30·6920 |
| 900 | 810000 | 30·0000 | 943 | 889249 | 30·7083 |
| 901 | 811801 | 30·0167 | 944 | 891136 | 30·7246 |
| 902 | 813604 | 30·0333 | 945 | 893025 | 30·7409 |
| 903 | 815409 | 30·0500 | 946 | 894916 | 30·7571 |

## A TABLE OF THE SQUARES AND SQUARE ROOTS OF NUMBERS.—Continued.

From 1 to 1000.

| No. | Squares. | Square Roots. | No. | Squares. | Square Roots. |
|---|---|---|---|---|---|
| 947 | 896809 | 30·7734 | 974 | 948676 | 31·2090 |
| 948 | 898704 | 30·7896 | 975 | 950625 | 31·2250 |
| 949 | 900601 | 30·8058 | 976 | 952576 | 31·2410 |
| 950 | 902500 | 30·8221 | 977 | 954529 | 31·2570 |
| 951 | 904401 | 30·8383 | 978 | 956484 | 31·2730 |
| 952 | 906304 | 30·8545 | 979 | 958441 | 31·2890 |
| 953 | 908209 | 30·8707 | 980 | 960400 | 31·3050 |
| 954 | 910116 | 30·8869 | 981 | 962361 | 31·3209 |
| 955 | 912025 | 30·9031 | 982 | 964324 | 31·3369 |
| 956 | 913936 | 30·9192 | 983 | 966289 | 31·3528 |
| 957 | 915849 | 30·9354 | 984 | 968256 | 31·3688 |
| 958 | 917764 | 30·9516 | 985 | 970225 | 31·3847 |
| 959 | 919681 | 30·9677 | 986 | 972196 | 31·4006 |
| 960 | 921600 | 30·9839 | 987 | 974169 | 31·4166 |
| 961 | 923521 | 31·0000 | 988 | 976144 | 31·4325 |
| 962 | 925444 | 31·0161 | 989 | 978121 | 31·4484 |
| 963 | 927369 | 31·0322 | 990 | 980100 | 31·4643 |
| 964 | 929296 | 31·0483 | 991 | 982081 | 31·4802 |
| 965 | 931225 | 31·0644 | 992 | 984064 | 31·4960 |
| 966 | 933156 | 31·0805 | 993 | 986049 | 31·5119 |
| 967 | 935089 | 31·0966 | 994 | 988036 | 31·5278 |
| 968 | 937024 | 31·1127 | 995 | 990025 | 31·5436 |
| 969 | 938961 | 31·1288 | 996 | 992016 | 31·5595 |
| 970 | 940900 | 31·1448 | 997 | 994009 | 31·5753 |
| 971 | 942841 | 31·1609 | 998 | 996004 | 31·5911 |
| 972 | 944784 | 31·1769 | 999 | 998001 | 31·6070 |
| 973 | 946729 | 31·1929 | 1000 | 1000000 | 31·6228 |

## TABLE OF SLOPES, &c.—For Topography.

| Degrees. | Vertical Rise in 100 feet horizontal. | Horizontal Distance to a rise of 10 feet. | Degrees. | Vertical Rise in 100 feet horizontal. | Horizontal Distance to a rise of 10 feet. |
|---|---|---|---|---|---|
| ° | | | ° | | |
| 1 | 1·75 | 572·9 | 19 | 34·43 | 29·0 |
| 2 | 3·49 | 286·4 | 20 | 36·40 | 27·5 |
| 3 | 5·24 | 190·8 | 21 | 38·40 | 26·0 |
| 4 | 6·99 | 143·0 | 22 | 40·40 | 24·7 |
| 5 | 8·75 | 114·3 | 23 | 42·45 | 23·5 |
| 6 | 10·51 | 95·1 | 24 | 44·52 | 22·4 |
| 7 | 12·28 | 81·4 | 25 | 46·63 | 21·4 |
| 8 | 14·05 | 71·2 | 26 | 48·77 | 20·5 |
| 9 | 15·83 | 63·1 | 27 | 50·95 | 19·6 |
| 10 | 17·63 | 56·7 | 28 | 53·17 | 18·8 |
| 11 | 19·44 | 51·4 | 29 | 55·43 | 18·0 |
| 12 | 21·25 | 47·0 | 30 | 57·73 | 17·3 |
| 13 | 23·09 | 43·3 | 35 | 70·02 | 14·2 |
| 14 | 24·93 | 40·1 | 40 | 83·91 | 11·9 |
| 15 | 26·79 | 37·3 | 45 | 100·00 | 10·0 |
| 16 | 28·67 | 34·9 | 50 | 119·17 | 8·4 |
| 17 | 30·57 | 32·7 | 55 | 142·81 | 7·0 |
| 18 | 32·49 | 30·7 | 60 | 173·20 | 5·7 |

E. B. MEARS, STEREOTYPER.

# CATALOGUE

OF

# PRACTICAL AND SCIENTIFIC BOOKS,

PUBLISHED BY

## HENRY CAREY BAIRD,

INDUSTRIAL PUBLISHER,

**No. 406 WALNUT STREET,**

**PHILADELPHIA.**

☞ Any of the Books comprised in this Catalogue will be sent by mail, free of postage, at the publication price.

☞ This Catalogue will be sent, free of postage, to any one who will furnish the publisher with his address.

**ARMENGAUD, AMOUROUX, AND JOHNSON.—THE PRACTICAL DRAUGHTSMAN'S BOOK OF INDUSTRIAL DESIGN, AND MACHINIST'S AND ENGINEER'S DRAWING COMPANION:** Forming a complete course of Mechanical Engineering and Architectural Drawing. From the French of M. Armengaud the elder, Prof. of Design in the Conservatoire of Arts and Industry, Paris, and MM. Armengaud the younger and Amouroux, Civil Engineers. Rewritten and arranged, with additional matter and plates, selections from and examples of the most useful and generally employed mechanism of the day. By WILLIAM JOHNSON, Assoc. Inst. C. E., Editor of "The Practical Mechanic's Journal." Illustrated by 50 folio steel plates and 50 wood-cuts. A new edition, 4to. . $10 00

**ARROWSMITH.—PAPER-HANGER'S COMPANION:** A Treatise in which the Practical Operations of the Trade are Systematically laid down: with Copious Directions Preparatory to Papering; Preventives against the Effect of Damp on Walls; the Various Cements and Pastes adapted to the Several Purposes of the Trade; Observations and Directions for the Panelling and Ornamenting of Rooms, &c. By JAMES ARROWSMITH, Author of "Analysis of Drapery," &c. 12mo., cloth . . . . . . . . . . $1 25

**BAIRD.—THE AMERICAN COTTON SPINNER, AND MANAGER'S AND CARDER'S GUIDE:**

A Practical Treatise on Cotton Spinning; giving the Dimensions and Speed of Machinery, Draught and Twist Calculations, etc.; with notices of recent Improvements: together with Rules and Examples for making changes in the sizes and numbers of Roving and Yarn. Compiled from the papers of the late ROBERT H. BAIRD. 12mo. . . . . $1 50

**BAKER.—LONG-SPAN RAILWAY BRIDGES:**

Comprising Investigations of the Comparative Theoretical and Practical Advantages of the various Adopted or Proposed Type Systems of Construction; with numerous Formulæ and Tables. By B. Baker. 12mo. . . . . . . $2 00

**BAKEWELL.—A MANUAL OF ELECTRICITY—PRACTICAL AND THEORETICAL:**

By F. C. BAKEWELL, Inventor of the Copying Telegraph. Second Edition. Revised and enlarged. Illustrated by numerous engravings. 12mo. Cloth . . . . . $2 00

**BEANS.—A TREATISE ON RAILROAD CURVES AND THE LOCATION OF RAILROADS:**

By E. W. BEANS, C. E. 12mo. (In press.)

**BLENKARN.—PRACTICAL SPECIFICATIONS OF WORKS EXECUTED IN ARCHITECTURE, CIVIL AND MECHANICAL ENGINEERING, AND IN ROAD MAKING AND SEWERING:**

To which are added a series of practically useful Agreements and Reports. By JOHN BLENKARN. Illustrated by fifteen large folding plates. 8vo. . . . . . . . $9 00

**BLINN.—A PRACTICAL WORKSHOP COMPANION FOR TIN, SHEET-IRON, AND COPPER-PLATE WORKERS:**

Containing Rules for Describing various kinds of Patterns used by Tin, Sheet-iron, and Copper-plate Workers; Practical Geometry; Mensuration of Surfaces and Solids; Tables of the Weight of Metals, Lead Pipe, etc.; Tables of Areas and Circumferences of Circles; Japans, Varnishes, Lackers, Cements, Compositions, etc. etc. By LEROY J. BLINN, Master Mechanic. With over One Hundred Illustrations. 12mo. $2 50

**BOOTH.—MARBLE WORKER'S MANUAL:**

Containing Practical Information respecting Marbles in general, their Cutting, Working, and Polishing; Veneering of Marble; Mosaics; Composition and Use of Artificial Marble, Stuccos, Cements, Receipts, Secrets, etc. etc. Translated from the French by M. L. Booth. With an Appendix concerning American Marbles. 12mo., cloth . . $1 50

**BOOTH AND MORFIT.—THE ENCYCLOPEDIA OF CHEMISTRY, PRACTICAL AND THEORETICAL:**

Embracing its application to the Arts, Metallurgy, Mineralogy, Geology, Medicine, and Pharmacy. By James C. Booth, Melter and Refiner in the United States Mint, Professor of Applied Chemistry in the Franklin Institute, etc., assisted by Campbell Morfit, author of "Chemical Manipulations," etc. Seventh edition. Complete in one volume, royal 8vo., 978 pages, with numerous wood-cuts and other illustrations. $5 00

**BOWDITCH.—ANALYSIS, TECHNICAL VALUATION, PURIFICATION, AND USE OF COAL GAS:**

By Rev. W. R. Bowditch. Illustrated with wood engravings. 8vo. . . . . . . . . . . $6 50

**BOX.—PRACTICAL HYDRAULICS:**

A Series of Rules and Tables for the use of Engineers, etc. By Thomas Box. 12mo. . . . . . . $2 00

**BUCKMASTER.—THE ELEMENTS OF MECHANICAL PHYSICS:**

By J. C. Buckmaster, late Student in the Government School of Mines; Certified Teacher of Science by the Department of Science and Art; Examiner in Chemistry and Physics in the Royal College of Preceptors; and late Lecturer in Chemistry and Physics of the Royal Polytechnic Institute. Illustrated with numerous engravings. In one vol. 12mo. . $2 00

**BULLOCK.—THE AMERICAN COTTAGE BUILDER:**

A Series of Designs, Plans, and Specifications, from $200 to to $20,000 for Homes for the People; together with Warming, Ventilation, Drainage, Painting, and Landscape Gardening. By John Bullock, Architect, Civil Engineer, Mechanician, and Editor of "The Rudiments of Architecture and Building," etc. Illustrated by 75 engravings. In one vol. 8vo. . . . . . . . . . . . $3 50

**BULLOCK.—THE RUDIMENTS OF ARCHITECTURE AND BUILDING:**

For the use of Architects, Builders, Draughtsmen, Machinists, Engineers, and Mechanics. Edited by JOHN BULLOCK, author of "The American Cottage Builder." Illustrated by 250 engravings. In one volume 8vo. . . . . $3 50

**BURGH.—PRACTICAL ILLUSTRATIONS OF LAND AND MARINE ENGINES:**

Showing in detail the Modern Improvements of High and Low Pressure, Surface Condensation, and Super-heating, together with Land and Marine Boilers. By N. P. BURGH, Engineer. Illustrated by twenty plates, double elephant folio, with text. $21 00

**BURGH.—PRACTICAL RULES FOR THE PROPORTIONS OF MODERN ENGINES AND BOILERS FOR LAND AND MARINE PURPOSES.**

By N. P. BURGH, Engineer. 12mo. . . . . $2 00

**BURGH.—THE SLIDE-VALVE PRACTICALLY CONSIDERED:**

By N. P. BURGH, author of "A Treatise on Sugar Machinery," "Practical Illustrations of Land and Marine Engines," "A Pocket-Book of Practical Rules for Designing Land and Marine Engines, Boilers," etc. etc. etc. Completely illustrated. 12mo. . . . . . . . . . . . $2 00

**BYRN.—THE COMPLETE PRACTICAL BREWER:**

Or, Plain, Accurate, and Thorough Instructions in the Art of Brewing Beer, Ale, Porter, including the Process of making Bavarian Beer, all the Small Beers, such as Root-beer, Ginger-pop, Sarsaparilla-beer, Mead, Spruce beer, etc. etc. Adapted to the use of Public Brewers and Private Families. By M. LA FAYETTE BYRN, M. D. With illustrations. 12mo. $1 25

**BYRN.—THE COMPLETE PRACTICAL DISTILLER:**

Comprising the most perfect and exact Theoretical and Practical Description of the Art of Distillation and Rectification; including all of the most recent improvements in distilling apparatus; instructions for preparing spirits from the numerous vegetables, fruits, etc.; directions for the distillation and preparation of all kinds of brandies and other spirits, spirituous and other compounds, etc. etc.; all of which is so simplified that it is adapted not only to the use of extensive distillers, but for every farmer, or others who may wish to engage in the art of distilling. By M. LA FAYETTE BYRN, M. D. With numerous engravings. In one volume, 12mo. $1 50

**BYRNE.—POCKET BOOK FOR RAILROAD AND CIVIL ENGINEERS:**

Containing New, Exact, and Concise Methods for Laying out Railroad Curves, Switches, Frog Angles and Crossings; the Staking out of work; Levelling; the Calculation of Cuttings; Embankments; Earth-work, etc. By OLIVER BYRNE. Illustrated, 18mo., full bound . . . . . . $1 50

**BYRNE.—THE HANDBOOK FOR THE ARTISAN, MECHANIC, AND ENGINEER:**

By OLIVER BYRNE. Illustrated by 11 large plates and 185 Wood Engravings. 8vo. . . . . . . . . $5 00

**BYRNE.—THE ESSENTIAL ELEMENTS OF PRACTICAL MECHANICS:**

For Engineering Students, based on the Principle of Work. By OLIVER BYRNE. Illustrated by Numerous Wood Engravings, 12mo. . . . . . . . . . $3 63

**BYRNE.—THE PRACTICAL METAL-WORKER'S ASSISTANT:**

Comprising Metallurgic Chemistry; the Arts of Working all Metals and Alloys; Forging of Iron and Steel; Hardening and Tempering; Melting and Mixing; Casting and Founding; Works in Sheet Metal; the Processes Dependent on the Ductility of the Metals; Soldering; and the most Improved Processes and Tools employed by Metal-Workers. With the Application of the Art of Electro-Metallurgy to Manufacturing Processes; collected from Original Sources, and from the Works of Holtzapffel, Bergeron, Leupold, Plumier, Napier, and others. By OLIVER BYRNE. A New, Revised, and improved Edition, with Additions by John Scoffern, M. B , William Clay, Wm. Fairbairn, F. R. S., and James Napier. With Five Hundred and Ninety-two Engravings; Illustrating every Branch of the Subject. In one volume, 8vo. 652 pages . $7 00

**BYRNE.—THE PRACTICAL MODEL CALCULATOR:**

For the Engineer, Mechanic, Manufacturer of Engine Work, Naval Architect, Miner, and Millwright. By OLIVER BYRNE. 1 volume, 8vo., nearly 600 pages . . . . $4 50

**CABINET MAKER'S ALBUM OF FURNITURE:**

Comprising a Collection of Designs for the Newest and Most Elegant Styles of Furniture. Illustrated by Forty eight Large and Beautifully Engraved Plates. In one volume, oblong $5 00

**CALVERT.—LECTURES ON COAL-TAR COLORS, AND ON RECENT IMPROVEMENTS AND PROGRESS IN DYEING AND CALICO PRINTING:**

Embodying Copious Notes taken at the last London International Exhibition, and *Illustrated with Numerous Patterns of Aniline and other Colors.* By F. GRACE CALVERT, F. R. S., F. C. S., Professor of Chemistry at the Royal Institution, Manchester, Corresponding Member of the Royal Academies of Turin and Rouen; of the Pharmaceutical Society of Paris; Société Industrielle de Mulhouse, etc. In one volume, 8vo., cloth . . . . . . . . . . . $1 50

**CAMPIN.—A PRACTICAL TREATISE ON MECHANICAL ENGINEERING:**

Comprising Metallurgy, Moulding, Casting, Forging, Tools, Workshop Machinery, Mechanical Manipulation, Manufacture of Steam-engines, etc. etc. With an Appendix on the Analysis of Iron and Iron Ores. By FRANCIS CAMPIN, C. E. To which are added, Observations on the Construction of Steam Boilers, and Remarks upon Furnaces used for Smoke Prevention; with a Chapter on Explosions. By R. Armstrong, C. E., and John Bourne. Rules for Calculating the Change Wheels for Screws on a Turning Lathe, and for a Wheel-cutting Machine. By J. LA NICCA. Management of Steel, including Forging, Hardening, Tempering, Annealing, Shrinking, and Expansion. And the Case-hardening of Iron. By G. EDE. 8vo. Illustrated with 29 plates and 100 wood engravings. $6 00

**CAMPIN.—THE PRACTICE OF HAND-TURNING IN WOOD, IVORY, SHELL, ETC.:**

With Instructions for Turning such works in Metal as may be required in the Practice of Turning Wood, Ivory, etc. Also, an Appendix on Ornamental Turning. By FRANCIS CAMPIN; with Numerous Illustrations, 12mo., cloth . . . $3 00

**CAPRON DE DOLE.—DUSSAUCE.—BLUES AND CARMINES OF INDIGO.**

A Practical Treatise on the Fabrication of every Commercial Product derived from Indigo. By FELICIEN CAPRON DE DOLE. Translated, with important additions, by Professor H. DUSSAUCE. 12mo. . . . . . . . . . $2 50

CAREY.—THE WORKS OF HENRY C. CAREY:

CONTRACTION OR EXPANSION? REPUDIATION OR RESUMPTION? Letters to Hon. Hugh McCulloch. 8vo. 38

FINANCIAL CRISES, their Causes and Effects. 8vo. paper 25

HARMONY OF INTERESTS; Agricultural, Manufacturing, and Commercial. 8vo., paper . . . . . . $1 00
Do. do. cloth . . . $1 50

LETTERS TO THE PRESIDENT OF THE UNITED STATES. Paper . . . . . . . . . . . . $1 00

MANUAL OF SOCIAL SCIENCE. Condensed from Carey's "Principles of Social Science." By KATE McKEAN. 1 vol. 12mo. . . . . . . . . . . . $2 25

MISCELLANEOUS WORKS: comprising "Harmony of Interests," "Money," "Letters to the President," "French and American Tariffs," "Financial Crises," "The Way to Outdo England without Fighting Her," "Resources of the Union," "The Public Debt," "Contraction or Expansion," "Review of the Decade 1857—'67," "Reconstruction," etc. etc. 1 vol. 8vo., cloth . . . . . . . . . . . $4 50

MONEY: A LECTURE before the N. Y. Geographical and Statistical Society. 8vo., paper . . . . . . 25

PAST, PRESENT, AND FUTURE. 8vo. . . . . $2 50

PRINCIPLES OF SOCIAL SCIENCE. 3 volumes 8vo., cloth $10 00

REVIEW OF THE DECADE 1857—'67. 8vo., paper 38

RECONSTRUCTION: INDUSTRIAL, FINANCIAL, AND POLITICAL. Letters to the Hon. Henry Wilson, U. S. S. 8vo. paper . . . . . . . . . . 38

THE PUBLIC DEBT, LOCAL AND NATIONAL. How to provide for its discharge while lessening the burden of Taxation. Letter to David A. Wells, Esq., U. S. Revenue Commission. 8vo., paper . . . . . . . . . 25

THE RESOURCES OF THE UNION. A Lecture read, Dec. 1865, before the American Geographical and Statistical Society, N. Y., and before the American Association for the Advancement of Social Science, Boston . . . 25

THE SLAVE TRADE, DOMESTIC AND FOREIGN; Why it Exists, and How it may be Extinguished. 12mo., cloth $1 50

THE WAY TO OUTDO ENGLAND WITHOUT FIGHTING HER. Letters to the Hon. Schuyler Colfax. 8vo., paper 75

**CAMUS.—A TREATISE ON THE TEETH OF WHEELS:**
Demonstrating the best forms which can be given to them for the purposes of Machinery, such as Mill-work and Clock-work. Translated from the French of M. CAMUS. By JOHN I. HAWKINS. Illustrated by 40 plates. 8vo. . . . . . . $3 00

**CLOUGH.—THE CONTRACTOR'S MANUAL AND BUILDER'S PRICE-BOOK:**
Designed to elucidate the method of ascertaining, correctly, the value and Quantity of every description of Work and Materials used in the Art of Building, from their Prime Cost in any part of the United States, collected from extensive experience and observation in Building and Designing; to which are added a large variety of Tables, Memoranda, etc., indispensable to all engaged or concerned in erecting buildings of any kind. By A. B. CLOUGH, Architect, 24mo., cloth 75

**COLBURN.—THE GAS-WORKS OF LONDON:**
Comprising a sketch of the Gas-works of the city, Process of Manufacture, Quantity Produced, Cost, Profit, etc. By ZERAH COLBURN. 8vo., cloth . . . . . . . 75

**COLBURN.—THE LOCOMOTIVE ENGINE:**
Including a Description of its Structure, Rules for Estimating its Capabilities, and Practical Observations on its Construction and Management. By ZERAH COLBURN. Illustrated. A new edition. 12mo. . . . . . . . . $1 25

**COLBURN AND MAW.—THE WATER-WORKS OF LONDON:**
Together with a Series of Articles on various other Waterworks. By ZERAH COLBURN and W. MAW. Reprinted from "Engineering." In one volume, 8vo. . . $4 00

**DAGUERREOTYPIST AND PHOTOGRAPHER'S COMPANION:**
12mo., cloth . . . . . . . . . . . $1 25

**DUPLAIS.—A COMPLETE TREATISE ON THE DISTILLATION AND PREPARATION OF ALCOHOLIC AND OTHER LIQUORS:**
From the French of M. DUPLAIS. Translated and Edited by M. McKENNIE, M. D. Illustrated. 8vo. (*In press.*)

**DESSOYE.—STEEL, ITS MANUFACTURE, PROPERTIES, AND USE.**

By J. B. J. Dessoye, Manufacturer of Steel; with an Introduction and Notes by Ed. Graten, Engineer of Mines. Translated from the French. In one volume, 12mo. (In press.)

**DIRCKS.—PERPETUAL MOTION:**

Or Search for Self-Motive Power during the 17th, 18th, and 19th centuries. Illustrated from various authentic sources in Papers, Essays, Letters, Paragraphs, and numerous Patent Specifications, with an Introductory Essay by Henry Dircks, C. E. Illustrated by numerous engravings of machines. 12mo., cloth . . . . . . . . . . $3 50

**DIXON.—THE PRACTICAL MILLWRIGHT'S AND ENGINEER'S GUIDE:**

Or Tables for Finding the Diameter and Power of Cogwheels; Diameter, Weight, and Power of Shafts; Diameter and Strength of Bolts, etc. etc. By Thomas Dixon. 12mo., cloth. $1 50

**DUNCAN.—PRACTICAL SURVEYOR'S GUIDE:**

Containing the necessary information to make any person, of common capacity, a finished land surveyor without the aid of a teacher. By Andrew Duncan. Illustrated. 12mo., cloth. $1 25

**DUSSAUCE.—A NEW AND COMPLETE TREATISE ON THE ARTS OF TANNING, CURRYING, AND LEATHER DRESSING:**

Comprising all the Discoveries and Improvements made in France, Great Britain, and the United States. Edited from Notes and Documents of Messrs. Sallerou, Grouvelle, Duval, Dessables, Labarraque, Payen, René, De Fontenelle, Malapeyre, etc. etc. By Prof. H. Dussauce, Chemist. Illustrated by 212 wood engravings. 8vo. . . . . . $10 00

**DUSSAUCE.—A GENERAL TREATISE ON THE MANUFACTURE OF EVERY DESCRIPTION OF SOAP:**

Comprising the Chemistry of the Art, with Remarks on Alkalies, Saponifiable Fatty Bodies, the apparatus necessary in a Soap Factory, Practical Instructions on the manufacture of the various kinds of Soap, the assay of Soaps, etc. etc. Edited from notes of Larmé, Fontenelle, Malapeyre, Dufour, and others, with large and important additions by Professor H. Dussauce, Chemist. Illustrated. In one volume, 8vo. $10 00

**DUSSAUCE.—A PRACTICAL GUIDE FOR THE PERFUMER:**
Being a New Treatise on Perfumery the most favorable to the Beauty without being injurious to the Health, comprising a Description of the substances used in Perfumery, the Formulæ of more than one thousand Preparations, such as Cosmetics, Perfumed Oils, Tooth Powders, Waters, Extracts, Tinctures, Infusions, Vinaigres, Essential Oils, Pastels, Creams, Soaps, and many new Hygienic Products not hitherto described. Edited from Notes and Documents of Messrs. Debay, Lunel, etc. With additions by Professor H. DUSSAUCE, Chemist. 12mo. press, *shortly to be issued.*) $3 00

**DUSSAUCE.—PRACTICAL TREATISE ON THE FABRICATION OF MATCHES, GUN COTTON, AND FULMINATING POWDERS.**
By Professor H. DUSSAUCE. 12mo. . . . . $3 00

**DUSSAUCE.—A GENERAL TREATISE ON THE MANUFACTURE OF VINEGAR, THEORETICAL AND PRACTICAL.**
Comprising the various methods, by the slow and the quick processes, with Alcohol, Wine, Grain, Cider, and Molasses, as well as the Fabrication of Wood Vinegar, etc. By Prof. H. DUSSAUCE. 12mo. (*In press.*)

**DE GRAFF.—THE GEOMETRICAL STAIR-BUILDERS' GUIDE:**
Being a Plain Practical System of Hand-Railing, embracing all its necessary Details, and Geometrically Illustrated by 22 Steel Engravings; together with the use of the most approved principles of Practical Geometry. By SIMON DE GRAFF, Architect. 4to. . . . . . . . . . . . . . $5 00

**DYER AND COLOR-MAKER'S COMPANION:**
Containing upwards of two hundred Receipts for making Colors, on the most approved principles, for all the various styles and fabrics now in existence; with the Scouring Process, and plain Directions for Preparing, Washing-off, and Finishing the Goods. In one vol. 12mo. . . . . . . $1 25

**EASTON.—A PRACTICAL TREATISE ON STREET OR HORSE-POWER RAILWAYS:**
Their Location, Construction, and Management; with General Plans and Rules for their Organization and Operation; together with Examinations as to their Comparative Advantages over the Omnibus System, and Inquiries as to their Value for Investment; including Copies of Municipal Ordinances relating thereto. By ALEXANDER EASTON, C. E. Illustrated by 23 plates, 8vo., cloth . . . . . . . . . $2 00

**FORSYTH.—BOOK OF DESIGNS FOR HEAD-STONES, MURAL, AND OTHER MONUMENTS:**

Containing 78 Elaborate and Exquisite Designs. By FORSYTH. 4to. (*In press.*)

**FAIRBAIRN.—THE PRINCIPLES OF MECHANISM AND MACHINERY OF TRANSMISSION:**

Comprising the Principles of Mechanism, Wheels, and Pulleys, Strength and Proportions of Shafts, Couplings of Shafts, and Engaging and Disengaging Gear. By WILLIAM FAIRBAIRN, Esq., C. E., LL. D., F. R. S., F. G. S., Corresponding Member of the National Institute of France, and of the Royal Academy of Turin; Chevalier of the Legion of Honor, etc. etc. Beautifully illustrated by over 150 wood-cuts. In one volume 12mo. $2 50

**FAIRBAIRN.—PRIME-MOVERS:**

Comprising the Accumulation of Water-power; the Construction of Water-wheels and Turbines; the Properties of Steam; the Varieties of Steam-engines and Boilers and Wind-mills. By WILLIAM FAIRBAIRN, C. E., LL. D., F. R. S., F. G. S. Author of "Principles of Mechanism and the Machinery of Transmission." With Numerous Illustrations. In one volume. (In press.)

**FLAMM.—A PRACTICAL GUIDE TO THE CONSTRUCTION OF ECONOMICAL HEATING APPLICATIONS FOR SOLID AND GASEOUS FUELS:**

With the Application of Concentrated Heat, and on Waste Heat, for the Use of Engineers, Architects, Stove and Furnace Makers, Manufacturers of Fire Brick, Zinc, Porcelain, Glass, Earthenware, Steel, Chemical Products, Sugar Refiners, Metallurgists, and all others employing Heat. By M. PIERRE FLAMM, Manufacturer. Illustrated. Translated from the French. One volume, 12mo. (In press.)

**GILBART.—A PRACTICAL TREATISE ON BANKING:**

By JAMES WILLIAM GILBART. To which is added: THE NATIONAL BANK ACT AS NOW (1868) IN FORCE. 8vo. $4 50

**GOTHIC ALBUM FOR CABINET MAKERS:**

Comprising a Collection of Designs for Gothic Furniture. Illustrated by twenty-three large and beautifully engraved plates. Oblong . . . . . . . . . $3 00

**GRANT.—BEET-ROOT SUGAR AND CULTIVATION OF THE BEET:**

By E. B. GRANT. 12mo. . . . . . . $1 25

**GREGORY.—MATHEMATICS FOR PRACTICAL MEN:**

Adapted to the Pursuits of Surveyors, Architects, Mechanics, and Civil Engineers. By OLINTHUS GREGORY. 8vo., plates, cloth . . . . . . . . . . $3 00

**GRISWOLD.—RAILROAD ENGINEER'S POCKET COMPANION.**

Comprising Rules for Calculating Deflection Distances and Angles, Tangential Distances and Angles, and all Necessary Tables for Engineers; also the art of Levelling from Preliminary Survey to the Construction of Railroads, intended Expressly for the Young Engineer, together with Numerous Valuable Rules and Examples. By W. GRISWOLD. 12mo., tucks. $1 50

**GUETTIER.—METALLIC ALLOYS:**

Being a Practical Guide to their Chemical and Physical Properties, their Preparation, Composition, and Uses. Translated from the French of A. GUETTIER, Engineer and Director of Founderies, author of "La Fouderie en France," etc. etc. By A. A. FESQUET, Chemist and Engineer. In one volume, 12mo. (In press, *shortly to be published.*)

**HATS AND FELTING:**

A Practical Treatise on their Manufacture. By a Practical Hatter. Illustrated by Drawings of Machinery, &c., 8vo. $1 25

**HAY.—THE INTERIOR DECORATOR:**

The Laws of Harmonious Coloring adapted to Interior Decorations: with a Practical Treatise on House-Painting. By D. R. HAY, House-Painter and Decorator. Illustrated by a Diagram of the Primary, Secondary, and Tertiary Colors. 12mo. $2 25

**HUGHES.—AMERICAN MILLER AND MILLWRIGHT'S ASSISTANT:**

By WM. CARTER HUGHES. A new edition. In one volume, 12mo. . . . . . . . . . $1 50

**HUNT.—THE PRACTICE OF PHOTOGRAPHY.**
By ROBERT HUNT, Vice-President of the Photographic Society, London, with numerous illustrations. 12mo., cloth 75

**HURST.—A HAND-BOOK FOR ARCHITECTURAL SURVEYORS:**
Comprising Formulæ useful in Designing Builder's work, Table of Weights, of the materials used in Building, Memoranda connected with Builders' work, Mensuration, the Practice of Builders' Measurement, Contracts of Labor, Valuation of Property, Summary of the Practice in Dilapidation, etc. etc. By J. F. HURST, C. E. 2d edition, pocket-book form, full bound $2 50

**JERVIS.—RAILWAY PROPERTY:**
A Treatise on the Construction and Management of Railways; designed to afford useful knowledge, in the popular style, to the holders of this class of property; as well as Railway Managers, Officers, and Agents. By JOHN B. JERVIS, late Chief Engineer of the Hudson River Railroad, Croton Aqueduct, &c. One vol. 12mo., cloth . . . . . . . $2 00

**JOHNSON.—A REPORT TO THE NAVY DEPARTMENT OF THE UNITED STATES ON AMERICAN COALS:**
Applicable to Steam Navigation and to other purposes. By WALTER R. JOHNSON. With numerous illustrations. 607 pp. 8vo., half morocco . . . . . . . . $6 00

**JOHNSON.—THE COAL TRADE OF BRITISH AMERICA:**
With Researches on the Characters and Practical Values of American and Foreign Coals. By WALTER R. JOHNSON, Civil and Mining Engineer and Chemist. 8vo. . . . $2 00

**JOHNSTON.—INSTRUCTIONS FOR THE ANALYSIS OF SOILS, LIMESTONES, AND MANURES.**
By J. W. F. JOHNSTON. 12mo. . . . . . 38

**KEENE.—A HAND-BOOK OF PRACTICAL GAUGING,**
For the Use of Beginners, to which is added A Chapter on Distillation, describing the process in operation at the Custom House for ascertaining the strength of wines. By JAMES B. KEENE, of H. M. Customs. 8vo. . . . . . $1 25

**KENTISH.—A TREATISE ON A BOX OF INSTRUMENTS,**
And the Slide Rule; with the Theory of Trigonometry and Logarithms, including Practical Geometry, Surveying, Measuring of Timber, Cask and Malt Gauging, Heights, and Distances. By THOMAS KENTISH. In one volume. 12mo. $1 25

**KOBELL.—ERNI.—MINERALOGY SIMPLIFIED:**

A short method of Determining and Classifying Minerals, by means of simple Chemical Experiments in the Wet Way. Translated from the last German Edition of F. VON KOBELL, with an Introduction to Blowpipe Analysis and other additions. By HENRI ERNI, M. D., Chief Chemist, Department of Agriculture, author of "Coal Oil and Petroleum." In one volume, 12mo. . . . . . . . . $2 50

**LAFFINEUR.—A PRACTICAL GUIDE TO HYDRAULICS FOR TOWN AND COUNTRY;**

Or a Complete Treatise on the Building of Conduits for Water for Cities, Towns, Farms, Country Residences, Workshops, etc. Comprising the means necessary for obtaining at all times abundant supplies of Drinkable Water. Translated from the French of M. JULES LAFFINEUR, C. E. Illustrated. (In press.)

**LAFFINEUR.—A TREATISE ON THE CONSTRUCTION OF WATER-WHEELS:**

Containing the various Systems in use with Practical Information on the Dimensions necessary for Shafts, Journals, Arms, etc., of Water-wheels, etc. etc. Translated from the French of M. JULES LAFFINEUR, C. E. Illustrated by numerous plates. (In press.)

**LANDRIN.—A TREATISE ON STEEL:**

Comprising the Theory, Metallurgy, Practical Working, Properties, and Use. Translated from the French of H. C. LANDRIN, Jr., C. E. By A. A. FESQUET, Chemist and Engineer. Illustrated. 12mo. $3 00

**LARKIN.—THE PRACTICAL BRASS AND IRON FOUNDER'S GUIDE:**

A Concise Treatise on Brass Founding, Moulding, the Metals and their Alloys, etc.; to which are added Recent Improvements in the Manufacture of Iron, Steel by the Bessemer Process, etc. etc. By JAMES LARKIN, late Conductor of the Brass Foundry Department in Reany, Neafie & Co.'s Penn Works, Philadelphia. Fifth edition, revised, with Extensive additions. In one volume, 12mo. . . . . . . $2 25

**LEAVITT.—FACTS ABOUT PEAT AS AN ARTICLE OF FUEL:**
With Remarks upon its Origin and Composition, the Localities in which it is found, the Methods of Preparation and Manufacture, and the various Uses to which it is applicable; together with many other matters of Practical and Scientific Interest. To which is added a chapter on the Utilization of Coal Dust with Peat for the Production of an Excellent Fuel at Moderate Cost, especially adapted for Steam Service. By H. T. LEAVITT. Third edition. 12mo. . . . . $1 75

**LEROUX.—A PRACTICAL TREATISE ON THE MANUFACTURE OF WORSTEDS AND CARDED YARNS:**
Translated from the French of CHARLES LEROUX, Mechanical Engineer, and Superintendent of a Spinning Mill. By Dr. H. PAINE, and A. A. FESQUET. Illustrated by 12 large plates. In one volume 8vo. . . . . . . . . . . $5 00

**LESLIE (MISS).—COMPLETE COOKERY:**
Directions for Cookery in its Various Branches. By MISS LESLIE. 58th thousand. Thoroughly revised, with the addition of New Receipts. In 1 vol. 12mo., cloth . . $1 25

**LESLIE (MISS). LADIES' HOUSE BOOK:**
a Manual of Domestic Economy. 20th revised edition. 12mo., cloth . . . . . . . . . . . . $1 25

**LESLIE (MISS).—TWO HUNDRED RECEIPTS IN FRENCH COOKERY.**
12mo. . . . . . . . . . . . 50

**LIEBER.—ASSAYER'S GUIDE:**
Or, Practical Directions to Assayers, Miners, and Smelters, for the Tests and Assays, by Heat and by Wet Processes, for the Ores of all the principal Metals, of Gold and Silver Coins and Alloys, and of Coal, etc. By OSCAR M. LIEBER. 12mo., cloth $1 25

**LOVE.—THE ART OF DYEING, CLEANING, SCOURING, AND FINISHING:**
On the most approved English and French methods; being Practical Instructions in Dyeing Silks, Woollens, and Cottons, Feathers, Chips, Straw, etc.; Scouring and Cleaning Bed and Window Curtains, Carpets, Rugs, etc.; French and English Cleaning, etc. By THOMAS LOVE. Second American Edition, to which are added General Instructions for the Use of Aniline Colors. 8vo. . . . . . . . . . . . 5 00

**MAIN AND BROWN.—QUESTIONS ON SUBJECTS CONNECTED WITH THE MARINE STEAM-ENGINE:**

And Examination Papers; with Hints for their Solution. By THOMAS J. MAIN, Professor of Mathematics, Royal Naval College, and THOMAS BROWN, Chief Engineer, R. N. 12mo., cloth $1 50

**MAIN AND BROWN.—THE INDICATOR AND DYNAMOMETER:**

With their Practical Applications to the Steam-Engine. By THOMAS J. MAIN, M. A. F. R., Ass't Prof. Royal Naval College, Portsmouth, and THOMAS BROWN, Assoc. Inst. C. E., Chief Engineer, R. N., attached to the R. N. College. Illustrated. From the Fourth London Edition. 8vo. . . . . $1 50

**MAIN AND BROWN.—THE MARINE STEAM-ENGINE.**

By THOMAS J. MAIN, F. R. Ass't S. Mathematical Professor at Royal Naval College, and THOMAS BROWN, Assoc. Inst. C. E. Chief Engineer, R. N. Attached to the Royal Naval College. Authors of "Questions connected with the Marine Steam-Engine," and the "Indicator and Dynamometer." With numerous Illustrations. In one volume, 8vo. . . . . $5 00

**MORTIMER.—THE PYROTECHNIST'S COMPANION:**

Or, a Familiar System of Recreative Fireworks. By G. W. MORTIMER. Illustrated 12mo. . . . . . . . $1 25

CONTENTS.—Introduction. Of Gunpowder, Materials, Apparatus, Division of Fireworks, Single Fireworks, Rockets, Tables of Various Compositions, Compound Fireworks.

**MARTIN.—SCREW-CUTTING TABLES, FOR THE USE OF MECHANICAL ENGINEERS:**

Showing the Proper Arrangement of Wheels for Cutting the Threads of Screws of any required Pitch; with a Table for Making the Universal Gas-Pipe Thread and Taps. By W. A. MARTIN, Engineer. 8vo. . . . . . . 50

**MILES.—A PLAIN TREATISE ON HORSE-SHOEING.**

With illustrations. By WILLIAM MILES, author of "The Horse's Foot," . . . . . . . . . $1 00

**MOLESWORTH. POCKET-BOOK OF USEFUL FORMULÆ AND MEMORANDA FOR CIVIL AND MECHANICAL ENGINEERS.**

By GUILFORD L. MOLESWORTH, Member of the Institution of Civil Engineers, Chief Resident Engineer of the Ceylon Railway. Second American, from the Tenth London Edition. In one volume, full bound in pocket-book form . . $2 00

**MOORE.—THE INVENTOR'S GUIDE:**
Patent Office and Patent Laws; or, a Guide to Inventors, and a Book of Reference for Judges, Lawyers, Magistrates, and others. By J. G. MOORE. 12mo., cloth . . $1 25

**NAPIER.—A SYSTEM OF CHEMISTRY APPLIED TO DYEING:**
By JAMES NAPIER, F. C. S. A New and Thoroughly Revised Edition, completely brought up to the present state of the Science, including the Chemistry of Coal Tar Colors. By A. A. FESQUET, Chemist and Engineer. With an Appendix on Dyeing and Calico Printing, as shown at the Paris Universal Exposition of 1867, from the Reports of the International Jury, etc. Illustrated. In one volume 8vo., 400 pages . . . . $5 00

**NAPIER.—A MANUAL OF DYEING RECEIPTS FOR GENERAL USE.**
By JAMES NAPIER, F. C. S. *With Numerous Patterns of Dyed Cloth and Silk.* Second edition, revised and enlarged. 12mo. $3 75

**NAPIER.—MANUAL OF ELECTRO-METALLURGY:**
Including the Application of the Art to Manufacturing Processes. By JAMES NAPIER. Fourth American, from the Fourth London edition, revised and enlarged. Illustrated by engravings. In one volume, 8vo. . . . . . $2 00

**NEWBERY.—GLEANINGS FROM ORNAMENTAL ART OF EVERY STYLE;**
Drawn from Examples in the British, South Kensington, Indian, Crystal Palace, and other Museums, the Exhibitions of 1851 and 1862, and the best English and Foreign works. In a series of one hundred exquisitely drawn Plates, containing many hundred examples. By ROBERT NEWBERY. 4to. $15 00

**NICHOLSON.—A MANUAL OF THE ART OF BOOK-BINDING:**
Containing full instructions in the different Branches of Forwarding, Gilding, and Finishing. Also, the Art of Marbling Book-edges and Paper. By JAMES B. NICHOLSON. Illustrated. 12mo., cloth . . . . . . . . $2 25

**NORRIS.—A HAND-BOOK FOR LOCOMOTIVE ENGINEERS AND MACHINISTS:**
Comprising the Proportions and Calculations for Constructing Locomotives; Manner of Setting Valves; Tables of Squares, Cubes, Areas, etc. etc. By SEPTIMUS NORRIS, Civil and Mechanical Engineer. New edition. Illustrated, 12mo., cloth $2 00

**NYSTROM.—ON TECHNOLOGICAL EDUCATION AND THE CONSTRUCTION OF SHIPS AND SCREW PROPELLERS:**

For Naval and Marine Engineers. By JOHN W. NYSTROM, late Acting Chief Engineer U. S. N. Second edition, revised with additional matter. Illustrated by seven engravings. 12mo. $2 50

**O'NEILL.—A DICTIONARY OF DYEING AND CALICO PRINTING:**

Containing a brief account of all the Substances and Processes in use in the Art of Dyeing and Printing Textile Fabrics; with Practical Receipts and Scientific Information. By CHARLES O'NEILL, Analytical Chemist; Fellow of the Chemical Society of London; Member of the Literary and Philosophical Society of Manchester; Author of "Chemistry of Calico Printing and Dyeing." To which is added An Essay on Coal Tar Colors and their Application to Dyeing and Calico Printing. By A. A. FESQUET, Chemist and Engineer. With an Appendix on Dyeing and Calico Printing, as shown at the Exposition of 1867, from the Reports of the International Jury, etc. In one volume 8vo., 491 pages. $6 00

**OVERMAN—OSBORN.—THE MANUFACTURE OF IRON IN ALL ITS BRANCHES:**

Including a Practical Description of the various Fuels and their Values, the Nature, Determination and Preparation of the Ore, the Erection and Management of Blast and other Furnaces, the characteristic results of Working by Charcoal, Coke, or Anthracite, the Conversion of the Crude into the various kinds of Wrought Iron, and the Methods adapted to this end. Also, a Description of Forge Hammers, Rolling Mills, Blast Engines, &c. &c. To which is added an Essay on the Manufacture of Steel. By FREDERICK OVERMAN, Mining Engineer. The whole thoroughly revised and enlarged, adapted to the latest Improvements and Discoveries, and the particular type of American Methods of Manufacture. With various new engravings illustrating the whole subject. By H. S. OSBORN, LL. D. Professor of Mining and Metallurgy in Lafayette College. In one volume, 8vo. $10 00

**PAINTER, GILDER, AND VARNISHER'S COMPANION:**

Containing Rules and Regulations in everything relating to the Arts of Painting, Gilding, Varnishing, and Glass Staining, with numerous useful and valuable Receipts; Tests for the Detection of Adulterations in Oils and Colors, and a statement of the Diseases and Accidents to which Painters, Gilders, and

Varnishers are particularly liable, with the simplest methods of Prevention and Remedy. With Directions for Graining. Marbling, Sign Writing, and Gilding on Glass. To which are added COMPLETE INSTRUCTIONS FOR COACH PAINTING AND VARNISHING. 12mo., cloth . . . . . . $1 50

**PALLETT.—THE MILLER'S, MILLWRIGHT'S, AND ENGINEER'S GUIDE.**

By HENRY PALLETT. Illustrated. In one vol. 12mo. $3 00

**PERKINS.—GAS AND VENTILATION.**

Practical Treatise on Gas and Ventilation. With Special Relation to Illuminating, Heating, and Cooking by Gas. Including Scientific Helps to Engineer-students and others. With illustrated Diagrams. By E. E. PERKINS. 12mo., cloth $1 25

**PERKINS AND STOWE.—A NEW GUIDE TO THE SHEET-IRON AND BOILER PLATE ROLLER:**

Containing a Series of Tables showing the Weight of Slabs and Piles to Produce Boiler Plates, and of the Weight of Piles and the Sizes of Bars to produce Sheet-iron; the Thickness of the Bar Gauge in Decimals; the Weight per foot, and the Thickness on the Bar or Wire Gauge of the fractional parts of an inch; the Weight per sheet, and the Thickness on the Wire Gauge of Sheet-iron of various dimensions to weigh 112 lbs. per bundle; and the conversion of Short Weight into Long Weight, and Long Weight into Short. Estimated and collected by G. H. PERKINS and J. G. STOWE . . . . $2 50

**PHILLIPS AND DARLINGTON.—RECORDS OF MINING AND METALLURGY:**

Or Facts and Memoranda for the use of the Mine Agent and Smelter. By J. ARTHUR PHILLIPS, Mining Engineer, Graduate of the Imperial School of Mines, France, etc., and JOHN DARLINGTON. Illustrated by numerous engravings. In one volume, 12mo. . . . . . . . . $2 00

**PRADAL, MALEPEYRE, AND DUSSAUCE.—A COMPLETE TREATISE ON PERFUMERY:**

Containing notices of the Raw Material used in the Art, and the Best Formulæ. According to the most approved Methods followed in France, England, and the United States. By M. P. PRADAL, Perfumer Chemist, and M. F. MALEPEYRE. Translated from the French, with extensive additions, by Professor H. DUSSAUCE. 8vo. . . . . . . . . $10 00

**PROTEAUX.—PRACTICAL GUIDE FOR THE MANUFACTURE OF PAPER AND BOARDS.**

By A. PROTEAUX, Civil Engineer, and Graduate of the School of Arts and Manufactures, Director of Thiers's Paper Mill, 'Puy-de-Dômé. With additions, by L. S. LE NORMAND. Translated from the French, with Notes, by HORATIO PAINE, A. B., M. D. To which is added a Chapter on the Manufacture of Paper from Wood in the United States, by HENRY T. BROWN, of the "American Artisan." Illustrated by six plates, containing Drawings of Raw Materials, Machinery, Plans of Paper-Mills, etc. etc. 8vo. . . . . . . . $5 00

**REGNAULT.—ELEMENTS OF CHEMISTRY.**

By M. V. REGNAULT. Translated from the French by T. FORREST BETTON, M. D., and edited, with notes, by JAMES C. BOOTH, Melter and Refiner U. S. Mint, and WM. L. FABER, Metallurgist and Mining Engineer. Illustrated by nearly 700 wood engravings. Comprising nearly 1500 pages. In two volumes, 8vo., cloth . . . . . . . $10 00

**REID.—A PRACTICAL TREATISE ON THE MANUFACTURE OF PORTLAND CEMENT:**

By HENRY REID, C. E. To which is added a Translation of M. A. Lipowitz's Work, describing a new method adopted in Germany of Manufacturing that Cement. By W. F. REID. Illustrated by plates and wood engravings. 8vo. . . . . . . $7 00

**SHUNK.—A PRACTICAL TREATISE ON RAILWAY CURVES AND LOCATION, FOR YOUNG ENGINEERS.**

By WM. F. SHUNK, Civil Engineer. 12mo. . . $1 50

**SMEATON.—BUILDER'S POCKET COMPANION:**

Containing the Elements of Building, Surveying, and Architecture; with Practical Rules and Instructions connected with the subject. By A. C. SMEATON, Civil Engineer, etc. In one volume, 12mo. . . . . . . . . . $1 50

**SMITH.—THE DYER'S INSTRUCTOR:**

Comprising Practical Instructions in the Art of Dyeing Silk, Cotton, Wool, and Worsted, and Woollen Goods: containing nearly 800 Receipts. To which is added a Treatise on the Art of Padding; and the Printing of Silk Warps, Skeins, and Handkerchiefs, and the various Mordants and Colors for the different styles of such work. By DAVID SMITH, Pattern Dyer. 12mo., cloth. . . . . . . . $3 00

**SMITH.—PARKS AND PLEASURE GROUNDS:**

Or Practical Notes on Country Residences, Villas, Public Parks, and Gardens. By CHARLES H. J. SMITH, Landscape Gardener and Garden Architect, etc. etc. 12mo. $2 25

**STOKES.—CABINET-MAKER'S AND UPHOLSTERER'S COMPANION:**

Comprising the Rudiments and Principles of Cabinet-making and Upholstery, with Familiar Instructions, Illustrated by Examples for attaining a Proficiency in the Art of Drawing, as applicable to Cabinet-work; The Processes of Veneering, Inlaying, and Buhl-work; the Art of Dyeing and Staining Wood, Bone, Tortoise Shell, etc. Directions for Lackering, Japanning, and Varnishing; to make French Polish; to prepare the Best Glues, Cements, and Compositions, and a number of Receipts particularly for workmen generally. By J. STOKES. In one vol. 12mo. With illustrations . . . . . $1 25

**STRENGTH AND OTHER PROPERTIES OF METALS.**

Reports of Experiments on the Strength and other Properties of Metals for Cannon. With a Description of the Machines for Testing Metals, and of the Classification of Cannon in service. By Officers of the Ordnance Department U. S. Army. By authority of the Secretary of War. Illustrated by 25 large steel plates. In 1 vol. quarto . . . . . . $10 00

**TABLES SHOWING THE WEIGHT OF ROUND, SQUARE, AND FLAT BAR IRON, STEEL, ETC.,**

By Measurement. Cloth . . . . . . . 63

**TAYLOR.—STATISTICS OF COAL:**

Including Mineral Bituminous Substances employed in Arts and Manufactures; with their Geographical, Geological, and Commercial Distribution and amount of Production and Consumption on the American Continent. With Incidental Statistics of the Iron Manufacture. By R. C. TAYLOR. Second edition, revised by S. S. HALDEMAN. Illustrated by five Maps and many wood engravings. 8vo., cloth . . . . $6 00

**TEMPLETON.—THE PRACTICAL EXAMINATOR ON STEAM AND THE STEAM-ENGINE:**

With Instructive References relative thereto, for the Use of Engineers, Students, and others. By WM. TEMPLETON, Engineer. 12mo. . . . . . . . . . $1 25

**THOMAS.—THE MODERN PRACTICE OF PHOTOGRAPHY.**
By R. W. THOMAS, F. C. S. 8vo., cloth . . . 75

**THOMSON.—FREIGHT CHARGES CALCULATOR.**
By ANDREW THOMSON, Freight Agent . . . . $1 25

**TURNBULL.—THE ELECTRO-MAGNETIC TELEGRAPH:**
With an Historical Account of its Rise, Progress, and Present Condition. Also, Practical Suggestions in regard to Insulation and Protection from the effects of Lightning. Together with an Appendix, containing several important Telegraphic Devices and Laws. By LAWRENCE TNRNBULL, M. D., Lecturer on Technical Chemistry at the Franklin Institute. Revised and improved. Illustrated. 8vo. . . . . $3 00

**TURNER'S (THE) COMPANION:**
Containing Instructions in Concentric, Elliptic, and Eccentric Turning; also various Plates of Chucks, Tools, and Instruments; and Directions for using the Eccentric Cutter, Drill, Vertical Cutter, and Circular Rest; with Patterns and Instructions for working them. A new edition in one vol. 12mo. $1 50

**ULRICH—DUSSAUCE.—A COMPLETE TREATISE ON THE ART OF DYEING COTTON AND WOOL:**
As practised in Paris, Rouen, Mulhausen, and Germany. From the French of M. LOUIS ULRICH, a Practical Dyer in the principal Manufactories of Paris, Rouen, Mulhausen, etc. etc.; to which are added the most important Receipts for Dyeing Wool, as practised in the Manufacture Impériale des Gobelins, Paris. By Professor H. DUSSAUCE. 12mo. $3 50

**URBIN—BRULL.—A PRACTICAL GUIDE FOR PUDDLING IRON AND STEEL.**
By ED. URBIN, Engineer of Arts and Manufactures. A Prize Essay read before the Association of Engineers, Graduate of the School of Mines, of Liege, Belgium, at the Meeting of 1865—6. To which is added a COMPARISON OF THE RESISTING PROPERTIES OF IRON AND STEEL. By A. BRULL. Translated from the French by A. A. FESQUET, Chemist and Engineer. In one volume, 8vo. . . . . . . . . $1 00

**WATSON.—A MANUAL OF THE HAND-LATHE.**
By EGBERT P. WATSON, Late of the "Scientific American," Author of "Modern Practice of American Machinists and Engineers." In one volume, 12mo. $1 50

**WATSON.—THE MODERN PRACTICE OF AMERICAN MACHINISTS AND ENGINEERS:**

Including the Construction, Application, and Use of Drills, Lathe Tools, Cutters for Boring Cylinders, and Hollow Work Generally, with the most Economical Speed of the same, the Results verified by Actual Practice at the Lathe, the Vice, and on the Floor. Together with Workshop management, Economy of Manufacture, the Steam-Engine, Boilers, Gears, Belting, etc. etc. By EGBERT P. WATSON, late of the "Scientific American." Illustrated by eighty-six engravings. 12mo. . . $2 50

**WATSON.—THE THEORY AND PRACTICE OF THE ART OF WEAVING BY HAND AND POWER:**

With Calculations and Tables for the use of those connected with the Trade. By JOHN WATSON, Manufacturer and Practical Machine Maker. Illustrated by large drawings of the best Power-Looms. 8vo. . . . . . . $10 00

**WEATHERLY.—TREATISE ON THE ART OF BOILING SUGAR, CRYSTALLIZING, LOZENGE-MAKING, COMFITS, GUM GOODS,**

And other processes for Confectionery, &c. In which are explained, in an easy and familiar manner, the various Methods of Manufacturing every description of Raw and Refined sugar Goods, as sold by Confectioners and others . . $2 00

**WILL.—TABLES FOR QUALITATIVE CHEMICAL ANALYSIS.**

By Prof. HEINRICH WILL, of Giessen, Germany. Seventh edition. Translated by CHARLES F. HIMES, Ph. D., Professor of Natural Science, Dickinson College, Carlisle, Pa. . $1 25

**WILLIAMS.—ON HEAT AND STEAM:**

Embracing New Views of Vaporization, Condensation, and Expansion. By CHARLES WYE WILLIAMS, A. I. C. E. Illustrated. 8vo. . . . . . . . . . . $3 50

**WOHLER.—A PRACTICAL TREATISE ON ANALYTICAL CHEMISTRY.**

By F. WÖHLER. With additions by GRANDEAU and TROOST. Edited by H. B. NASON, Professor of Chemistry, Rensselaer Institute, Troy, N. Y. With numerous Illustrations. (*In press.*)

**WORSSAM.—ON MECHANICAL SAWS:**

From the Transactions of the Society of Engineers, 1867. By S. W. WORSSAM, Jr. Illustrated by 18 large folding plates. 8vo. $5 00

**BOX.—A PRACTICAL TREATISE ON HEAT AS APPLIED TO THE USEFUL ARTS:**

For the use of Engineers, Architects, etc. By THOMAS BOX, author of "Practical Hydraulics." Illustrated by 14 plates, containing 114 figures. 12mo. . . . . . . . $4 25

**BYRNE.—THE AMERICAN ENGINEER, DRAUGHTSMAN, AND MACHINIST'S ASSISTANT:**

Designed for Practical Workingmen, Apprentices, and those intended for the Engineering Profession. Illustrated with 200 Engravings on wood, and 14 large Plates of American Machinery and Engine-work. By OLIVER BYRNE. 4to. Cloth $6 00

**CHAPMAN.—A TREATISE ON ROPE-MAKING,**

As practised in private and public Rope-yards, with a Description of the Manufacture, Rules, Tables of Weights, etc. adapted to the Trade; Shipping, Mining, Railways, Builders, etc. By ROBERT CHAPMAN. 24mo. . . . . . . . . . . $1 50

**SLOAN.—AMERICAN HOUSES:**

A variety of Original Designs for Rural Buildings. Illustrated by 26 colored Engravings, with Descriptive References By SAMUEL SLOAN, Architect; author of the "Model Architect," etc. etc. 8vo $2 50

**SMITH.—THE PRACTICAL DYER'S GUIDE:**

Comprising Practical Instructions in the Dyeing of Shot Cobourgs, Silk Striped Orleans, Colored Orleans from Black Warps, ditto from White Warps, Colored Cobourgs from White Warps, Merinos, Yarns, Woollen Cloths, etc. Containing nearly 300 Receipts, to most of which a Dyed Pattern is annexed. Also, a Treatise on the Art of Padding. By DAVID SMITH. In one vol. 8vo. $25 00

www.ingramcontent.com/pod-product-compliance
Lightning Source LLC
LaVergne TN
LVHW021421110826
845150LV00007B/2019

* 9 7 8 1 4 2 5 5 0 9 8 2 8 *